D1187544

A Graphical Engineering Aid for VLSI Systems

Computer Science: Computer Architecture and Design, No. 4

Harold S. Stone, Series Editor

IBM Corporation
T.J. Watson Research Center
Yorktown, New York

Other Titles in This Series

No. 1	*Automated Microcode Synthesis*	Robert A. Mueller
No. 2	*A Parallel Pipeline Computer Architecture for Speech Processing*	Vassilios John Georgiou
No. 3	*Algorithms for Some Design Automation Problems*	James P. Cohoon

A Graphical Engineering Aid for VLSI Systems

by
Paul J. Drongowski
A.R. Jennings Distinguished Assistant Professor
Department of Computer Engineering and Science
Case Western Reserve University
Cleveland, Ohio

Produced and distributed by
UMI Research Press
an imprint of
University Microfilms International
A Xerox Information Resources Company
Ann Arbor, Michigan 48106

Library of Congress Cataloging in Publication Data

Drongowski, Paul J., 1952-
A graphical engineering aid for VLSI systems.

(Computer science. Computer architecture and design ; no. 4)
Revision of thesis (Ph.D.)—University of Utah, 1982.
Bibliography: p.
Includes index.
1. Integrated circuits—Very large scale integration—Design and construction. 2. Computer-aided design.
I. Title. II. Title: Graphical engineering aid for V.L.S.I. systems. III. Series.
TK7874.D76 1985 621.3819'5835 85-1041
ISBN 0-8357-1656-2 (alk. paper)

For my Mom and Dad,
Helen and Sigmont Drongowski

Contents

Preface

The effect of very large scale integration (VLSI) on consumer, industrial, and military electronics is pervasive. The reduced size and increased power of VLSI computing engines have encouraged the development of at least one new industry (personal computers) and revolutionized others, from industrial process control to sewing machines and microwave ovens. VLSI systems are particularly well suited for high volume applications where increased functionality at reduced cost provides a key competitive advantage.

For a product to succeed in the marketplace, it must satisfy a particular demand when the demand arises at a reasonable price. A product that is late, expensive, or inappropriate simply will fail. Computer aided design (CAD) tools can help an engineering team to exploit VLSI technology more effectively by:

—Decreasing development time (getting the product to market sooner),

—Reducing design complexity (producing a more reliable and useable product), and

—Decreasing nonrecurring engineering costs (lowering the overall price of the product).

Contemporary engineering workstations and CAD tools are beginning to fill these needs through schematic capture, simulation, layout, and analysis programs.

This book describes an engineering aid for the design of VLSI systems which emphasizes a functional (or linguistic) approach to the description, analysis, and synthesis of VLSI circuits. Tool builders and researchers alike should find several aspects of this work novel and useful. The specification and use of a graphical hardware description language (HDL) based upon representational datatypes is a significant departure from textual (or symbolic) HDLs. Algorithms that map a design to its implementation and that calculate the expected execution time of a system are given in the appendices and should provide a basis for future work or application. For students who are interested in digital systems CAD, this book will present a comprehensive view of an experimental CAD system. The interrelationships between notation, design databases, and tools are explored. More recent results and work are reported in the epilogue.

Special thanks go to Al Davis (Fairchild Flair), Chuck Rose (Case Western Reserve and Endot Inc.), and Sam Fuller (Digital Equipment Corp.) for their long-term (and sometimes long-distance) guidance and support. Al's advice, encouragement, and technical expertise were invaluable. Since Chuck and Sam helped me to form many of my early impressions of CAD and computer architecture, they, too, have been quite influential.

I would like to thank the following friends and colleagues for their constructive comments and insights: Gary Lindstrom, Bob Keller, P. Subrahmanyam, Kent Smith, Elliott Organick, Lino Ferretti, Uri Weiser, Dave Matty, Larry Seppi, and Ed Snow.

I will always be grateful for the cheering section in Cleveland: my parents, Lydia, Jack, Alicia, and Carl Boldt. Finally, I would like to thank Francie Hunt, whose unfailing patience and encouragement got me through it all then—and now.

1

Introduction

High density integrated circuitry has offered an unprecedented opportunity to extend the capabilities of modern computing tools. The design of systems based on the technology of very large scale integration (VLSI) also poses some unprecedented challenges [Sut77].

Historical evidence indicates that chip complexity has grown exponentially since about 1971. Foreseeable improvements in fabrication technology guarantee a continuing increase in circuit densities at this phenomenal rate. A recent study conducted at Intel Corporation shows that designer productivity has declined for the last several years while complexity has increased [Lat79]. In 1982, it took an estimated 60 person-years just to lay out the design of a microcomputer chip. Clearly, our computer-based tools must be drastically improved to gain increased leverage on the integrated circuit design problem.

This book presents and evaluates a method for the design of nMOS VLSI systems. This method, called d, is a discipline supporting top-down and bottom-up hierarchical design. The system to be designed is described in a graphical notation, d-n, which separates the specification of system behavior and structure from its physical design, moving VLSI design from a preoccupation with geometry toward the realm of programming. Through floor plan diagrams and mechanical delay analysis, the designer can stay in touch with the real physical concerns of space utilization and execution speed. Some experimental software tools were constructed to demonstrate the feasibility of a real, complete computer aided design (CAD) system using this method.

N-channel MOS was selected as the subject technology in this research for several reasons:

1. nMOS has sufficient component density to support the design of very complex systems, the topic of this study.

2. It is a popular technology and will be used extensively through 1985. Results from this work, therefore, would have fairly wide and timely impact on the IC design community.

3. Expertise on nMOS circuit design and device behavior is readily available.

4. The local computing environment includes some design tools for nMOS circuits, making software experiments a bit easier to conduct.

5. Other scientists are proposing schemes for the design of nMOS systems. By using the same technology, objective comparisons between methods and results can be made.

In this book, the term "VLSI" will mean "nMOS VLSI."

Chapter 2 examines some of the major issues affecting the design of VLSI systems. It presents the conventional approach to VLSI design and looks at contemporary efforts to improve our tools and techniques.

In chapter 3, the goals and guiding principles of the d methodology are outlined. The notation for d-based VLSI designs, d-n, is presented in chapter 4. A detailed, reference-style presentation of d-n is made in appendix A. The tools to support the language and the methodology are proposed in chapter 5.

The design of a cache memory system in d-n is presented in chapter 6. Chapter 7 discusses the experimental tools which were developed as part of this project and makes some suggestions for the future implementation of d. Because a tool both supports and constrains its user, the relationship between d and the VLSI designer is an important one. This subject is explored in chapter 8.

Chapter 9 states the critical conclusions which may be drawn from this work. It also presents myriad possibilities for future research and extensions.

2

VLSI Design

Background

Contemporary disciplines for system design encourage the use of both top-down and bottom-up methods. In top-down design, overall system function is successively decomposed into smaller, more manageable units. Bottom-up design is a process of synthesis in which building blocks are combined to form still larger building blocks. If the resulting system is structured such that lower level elements are not defined in terms of higher level elements, then the system is said to be hierarchical.

A design team usually begins work with an initial specification that describes the intended behavior of the system. Subsequent design activities may be separated into three domains. Behavioral design decomposes and refines the initial specification into a hierarchy of subproblems. These subproblems and their solutions are assigned to individual modules or subsystems. Structural design involves the interconnection of these modules to form buildings which may be interconnected at still higher levels in the hierarchy. Often behavioral and structural design are performed in concert. Eventually the behavioral and structural designs must be implemented in a particular technology. The actual representation of the behavior and structure of the system in this technology is the object of physical design [Van79].

Often the initial specification is found to be ambiguous or incomplete, resulting in some redesign. Changes in the initial specification will directly affect the behavioral design and indirectly, the structural design. For example, the system structural and behavioral specification may be inadequate to support some new operational requirement which was added to the specification by management after design had already started. In other cases, redesign may result from unanticipated interactions between domains. For example, structural design normally terminates when the engineers feel that each of the building blocks at the bottom of the hierarchy is "sufficiently primitive" to be directly implemented in the chosen technology. If this is not the case, structural design must be continued. This is an example of interaction between the structural and physical domains.

General considerations aside, the design of a VLSI-based digital system brings along its own set of problems, largely due to the nature of the implementation technology itself. (The remainder of this section is an overview of the nMOS process and its physical implications for digital system design.) The ultimate product of the VLSI design process is a set of masks which are used to pattern successive layers of circuit materials on a silicon substrate. The physical process of forming the patterns and layers is called fabrication. The choice of materials, layers, and patterns depends upon the kinds of circuit devices to be employed and the fabrication process to be used to manufacture the chips. For example, the popular n-channel MOS (nMOS) technology consists of four layers of materials (n-channel silicon, polysilicon, aluminum metal, and p-bulk) separated by insulating layers of silicon dioxide. Any of these conducting paths may cross and openings may be formed in the insulating layers to jump the circuit between layers. Wherever polysilicon (or just "poly" for short) crosses an n-channel region, an nMOS transistor is formed. These transistors are the basic active circuit elements and may be interconnected by two and one-half layers of wiring.[1] The masks produced by the VLSI design process determine the physical position and size of the transistors and wires on the chip—the so-called "chip geometry."

Since the transistors and wires are composed of semiconducting materials their electrical behavior follows the natural principles of device physics. Polysilicon, n-channel, and metal wires act as capacitors and resistors. Hence, circuits which drive these wires are charging minute capacitors and are incurring signal delays. The resistivity of the wire path determines the amount of signal drop across the wire, affecting noise immunity and logic thresholds. Both the resistance and capacitance of a wire are proportional to its length and depend upon the wire material.

Behavior of the nMOS transistor is determined by the physical dimensions of the crossing n-channel and poly paths. The amount of current which a transistor can source or sink depends upon the length and width of the transistor gate (the intersection of the poly and n-channel region). The capacitance of the nMOS transistor gate is a very significant factor in circuit delays and is also determined by the transistor geometry.

On-chip capacitances rarely exceed 1 picofarad and, therefore, make modest current demands on most of the circuitry. When a signal must be driven off the chip, however, the capacitive load increases to approximately 100 picofarads. Therefore, on-chip signals can be switched 100 times as fast as off-chip signals because delay is roughly proportional to load capacitance. In addition, much larger transistors and multistage amplifiers must be used to drive the signal into the bigger load. Hence, there is a space (smaller driver size) and time (shorter signal delay) advantage to be gained by packing as much functionality as possible onto a single chip [Dav79].

Signal delays affect system timing. Systems that are fully synchronous depend upon a global clock to sequence control events. In order to drive the clock signal into the relatively long control wires that pervade every region of the chip without excessive delay, a large clock generator and buffer must be used. In a highly complex digital system such as a VLSI circuit, it is hard to guarantee that the clock period will be sufficiently long to prevent hazards due to signal skew. There are simply too many assumptions about timing behavior to be completely verified.

A more practical approach is the self-timed system concept of Seitz [Sei79, Sei80]. In this kind of organization, the separate subsystems are responsible for keeping their own notion of time (e.g., a local clock). Assumptions about synchronous behavior, therefore, are localized at the subsytem level and communication between subsystems is performed asynchronously. This eliminates errors due to signal skew.

Transistors and wires cannot be placed on the chip haphazardly. Due to certain limitations in the fabrication process, physical device dimensions are restricted by minumum size and spacing constraints called design rules. Sizes and separations are specified as multiples of the minimum feature size. The feature size is largely determined by the accuracy of the etching, implantation, and alignment steps in the fabrication process. Some of the design rules have their basis in device electronics. For example, an unrelated (unconnected) poly wire must not overlap a parallel n-channel diffusion wire or an unwanted transistor will be created. It is not an easy task to guarantee adherence to the design rules when a system has 500,000 rectangles such as a chip in the Intel iAPX 432 family [Lat79].

Not all the chips on a particular wafer will operate correctly after fabrication. The percentage of working chips is commonly referred to as the yield. Most chip failures are due to defects in the silicon wafer, mask flaws, or over (under) etching the layers. One model for estimating chip failure assumes that flaws are uniformly and randomly distributed across the wafer surface. By minimizing chip area, yield can be improved for a given design [Mea80]. Additional failures can be expected after packaging as a result of bonding errors between the chip and the wire carrying the external signal to the package pin. By keeping the number of pins low, the overall yield (after packaging) can be increased. Clearly, chip area and yield must be balanced against functionality and the reduction of off-chip communication.

From the preceding discussion, the physical nature of VLSI systems can be seen to be a very heavy influence on the behavioral and structural design of these systems. During design, the initial specification must be taken from a very high level description of system behavior to the actual geometry of transistors and wires. Wire lengths and transistor sizes directly affect the execution speed of the system, the amount of functionality which can be squeezed onto the chip (density), and even the yield which can be expected after fabrication and packaging. Structural design must accommodate not only the placement of modules, but also the routing of wires between modules. Wire length must be kept to a reasonable distance to prevent signal delays from becoming too long or the size of drive transistors from becoming too large. In addition to system level issues, there is the matter of managing all the details which must "play together" in order to form a working system. Design rules must be followed and transistor sizes chosen appropriately. All interconnections between subsystems must be made correctly. Clearly, VLSI design is an exercise in complexity management—a problem begging for a structured design discipline.

Current Practice

Existing VLSI design methods and tools are heavily oriented toward the generation of artwork (fabrication masks) and the validation of system design through simulation. Figure 1 illustrates the typical industrial design process.

Figure 1. VLSI Design Process as Practiced by Industry

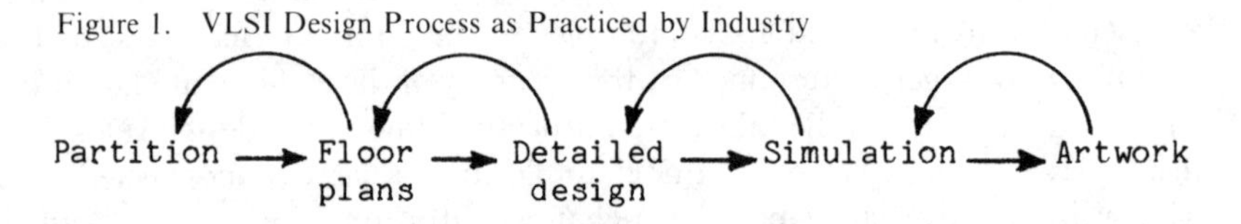

The design team begins with an initial specification and decomposes the problem into subproblems. During this stage of design, the system structure is drawn as a set of block diagrams showing the major subsystems, their interfaces, and the interconnections between the subsystems. Frequently, the total design problem is too big to fit on a single circuit chip. The decomposition process must, therefore, take subsystem sizes and off-chip communication times into account. Decomposition is usually followed by a sizing step in which the physical dimension of each subsystem is estimated according to its function. If a group of subsystems cannot fit on a chip, they must be subdivided further. If subdivision is impossible, the chip design must be abandoned!

Once a suitable structure has been found, the block diagram is translated into a floor plan indicating the relative position and size of each subsystem. The detailed design of the subsystems is then started. First, the major subsystem components such as programmable logic arrays (PLA), read only memories (ROM), random access memories (RAM), registers, and arithmetic units are identified. Preliminary subsystem layouts are drawn leaving room for the interconnections, control logic, and provisions for circuit testing. The preliminary subsystem layouts are compared with the floor plan. If the initial size estimates are wrong, and they usually are, the floor plan must be rearranged to accommodate the changes [SmiPC].

Computer graphics has been a valuable tool for planning and detailed design. Revisions can be made rapidly and the designer quickly discovers what will and will not work. Many of these systems incorporate automatic design rule checking (DRC) to catch "syntax" errors as early as possible.

After a subsystem has been fully detailed, its operation can be simulated. Due to the amount of computing resources required, electrical simulation cannot be used to analyze circuits with more than 20 to 30 transistors. The design must be translated to a gate or register transfer (RT) level description to be simulated as a complete system since the ratio of real to simulated time is much lower for this kind of simulation.

With gate or RT level simulation, information about system timing is sometimes lost. Using the gate and register delays which have been determined through electrical simulation, the individual circuit delays are combined into an aggregate delay for the subsystem. This is done manually with all the attendant problems of human fallibility. Further, this strategy does not account for delays due to wire capacitance. If either the simulation results or the delay analysis are unsatisfactory, the detailed subsystem design must be changed. Eventually, a satisfactory design is found, the layout is finalized, the design rules are checked one last time, and pattern generation (PG) tapes are produced from the circuit geometry. The PG tapes are then used to make the glass circuit masks for the fabrication process.

Contemporary Approaches

This section describes contemporary research efforts to produce better VLSI design tools.

Caltech

The VLSI research at Caltech is funded by a consortium of industrial organizations and is perhaps the largest of the university efforts.

Much of the Caltech work has been inspired by Carver Mead. In their popular book, Mead and Lynn Conway have proposed a method for VLSI design that combines top-down design with a notation and technique for the design of low to medium complexity, special purpose circuit cells [Mea80]. Using this method, the engineer decomposes the system into modules from which a floor plan for the chip is drawn. The modules consist primarily of highly regular circuit structures such as PLAs, RAMs, ROMs, and special purpose circuit cells that are replicated many times across the surface of the chip. Complexity is reduced through the use of regular structures because:

—Once a cell and its interactions with its neighbors are understood, the cell can be safely repeated;

—Replication of a carefully and cleverly laid out cell suppresses the complexity of the local wiring.

The special purpose cells are designed using the "stick" notation. First, a schematic for the circuit is drawn showing all nMOS transistors and their interconnections. A rough layout of the circuit is made from its schematic ignoring design rules and transistor geometries. It is composed of colored sticks where each stick represents a circuit path and the stick color indicates the layer where the path will appear on the chip. Contacts and implantation regions are also indicated.

Stick diagrams show the general topology of a circuit, permitting the designer to compact the layout without immediate concern for the design rules. The transistor sizes are chosen through electrical simulation or through rules of thumb (design by ratio). This choice is itself a difficult problem:

—The circuit delay and gain are determined by the size ratio of the pull-up transistors to the pull-down transistors in an *isolated* nMOS circuit.

—When the same circuit is connected to the rest of the system, its operation will be affected by the surrounding electrical network. Adjustments to the transistor sizes in the circuit *and the electrical network* may be required. Electrical design cannot be easily isolated.

After the transistor sizes have been selected, the designer uses the rough topography of the stick diagram and the design rules to draw the actual circuit layout.

At the architectural level, system control is implemented by a microprogrammable controller using a PLA or ROM. At the timing level, all combinatorial (data) operations are assigned to the first phase of a two-phase, nonoverlapping clock, while register transfer operations are assigned to the second phase. The clock period must be as long as the maximum combinatorial delay in the circuit unless a more complex, programmable clock is used instead. With such a discipline, the system behaves like a classical register transfer machine.

What limits the method of Mead and Conway is its lack of formality. The top-down technique is an informal prescription for system decomposition and does not provide a formal language with well-specified semantics. Without a formal language and its associated mathematical properties, mechanical translation and analytical aids cannot be built. The sticks form a convenient language for sketching low complexity circuits but do not provide much assistance for subsystem level design. A standard has been proposed for this notation and it has been used in at least two cell geometry editors [Mas81, Tri80, Wil77].

"Bristle Blocks" is a technique for detailed design [Joh79]. The goals of this method are to hide the mechanical aspects of cell layout and to provide for a high degree of optimization in the layout. The use of highly regular structures and hierarchical design is encouraged as in the method of Mead and Conway.

The fundamental unit in Bristle Blocks is the "cell" (fig. 2). Each cell is composed of geometric primitives (e.g., rectangles), references to other cells, and connection points that extend from the cell like bristles. Cell layout is left largely to the ingenuity of the designer. Bristle Blocks depend upon a rigid physical format (fig. 3) which consists of a "core" of information processing elements that are controlled by an "instruction decoder" and intercommunicate over two busses. The logic is driven by a two-phase, nonoverlapping clock with register transfers assigned to Phi-1 and data operations to Phi-2. All the core elements share a common width, called pitch. Interconnections are spaced and layered along the common sides such that the elements "plug" together. This technique reduces the amount of local wiring which must be made including power distribution.

Figure 2. A Bristle Block Cell

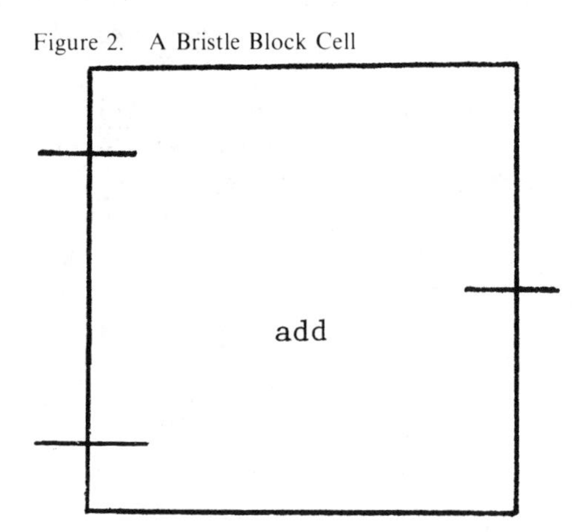

Bristle Blocks have several drawbacks. All core elements must be made as wide as the widest cell, whose pitch is not known until all the cells have been designed. Stretchable cells, which can deform to the appropriate pitch, have been suggested as a solution to this problem. However, cell design is further

Figure 3. Physical Format of a Bristle Block System

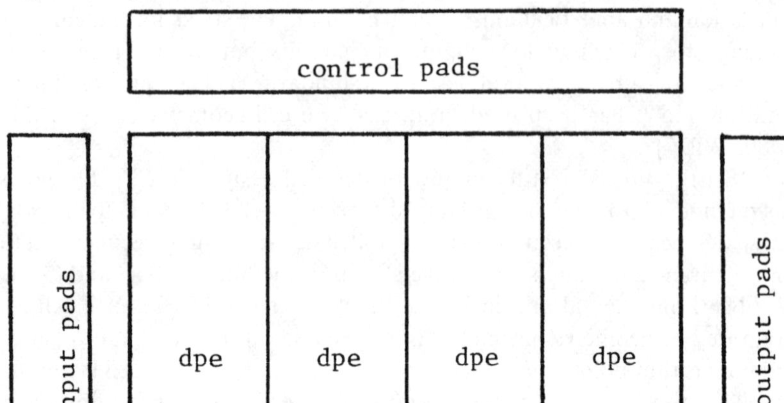

complicated by shifting the local wiring problem to clever cell layout. Finally, designs are restricted to a particular organization—systems that can be supported by the two-bus, global clock scheme. This imposes a serious constraint upon the kinds of architectures which can be implemented in Bristle Blocks.

"Blocks" are a more general approach to the VLSI design problem than Bristle Blocks although the techniques have common features [GraCTb]. Systems are made of blocks which may be as simple as a transistor or as complex as an adder or a subsystem. Blocks may be interconnected between "connection points" forming a structural specification of the system. Since blocks have behavioral, structural, and physical properties, their descriptions can be used for simulation and design analysis in addition to the production of the circuit geometry.

The notion of a "coordinode" is crucial to the formation of block specifications. Coordinodes correspond to connection points; coinciding coordinodes represent a junction between two components. Some coordinodes may be specified as "pins" (external connections) and can be used to provide block instance connections. Behaviorally, coordinodes belong to an electrical network providing information that can be exploited for low level simulation.

Figure 4 shows the structural and physical definitions of a shift register cell from [GraCtb]. The physical definition is derived from the structural one by instantiating each of the subblocks and wires with a spatial transformation

Figure 4. Coordinode Structural and Physical Definitions

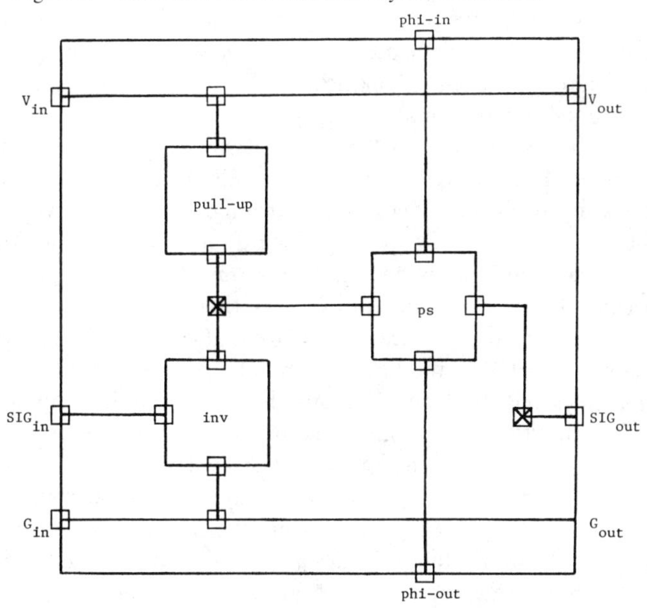

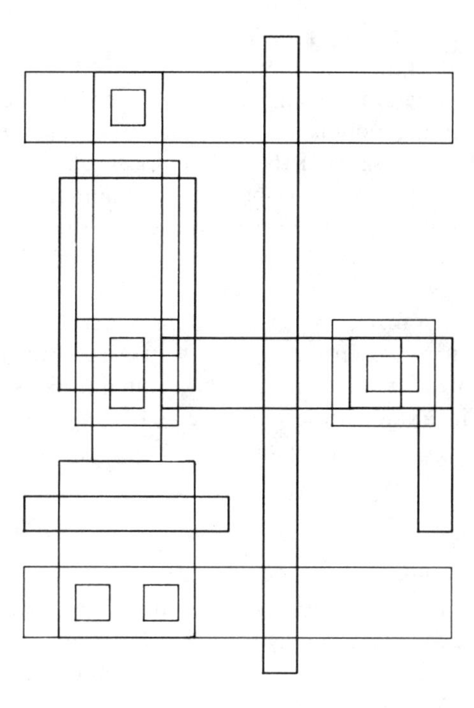

(e.g., translation, rotation, mirroring) and a set of parameters. These parameters may control cell deformation, array dimensions, programmability, and conditional inclusion of circuit elements. The procedural representation of a block is a real boon to adaptability, making a single definition applicable to a wider variety of situations than a static specification.

Due to the rigid enforcement of hierarchy, wiring must be completed incrementally. It is not clear how blocks would enable the designer to make good strategic layout decisions. The blocks technique relies heavily on simulation for both validation and performance analysis. The cycle from design to performance evaluation to redesign will be quite lengthy and perhaps unacceptable because large scale simulation is very time consuming.

Behavioral, structural, and physical design are unified in blocks and they share the same notation. Hence, the behavioral and structural aspects of the design are still cluttered with extraneous physical detail, impeding comprehension of the logical design. The position and size of block components directly reflect the placement and size of physical elements. Information of this kind in the structural diagram may encourage designers to optimize the low level design too early. From the software engineering experience, early optimization extends the design period because optimization effort is sometimes thrown away due to revocation or redefinition of higher level decisions.

In a related effort at Caltech, James Rowson used mathematical combinators and functions to model hierarchically organized VLSI systems. (A combinator is a mathematical operator which maps one or more functions into a new function and is gaining popularity as a basis for functional programming languages [Tur79].) In this model, the hierarchy is separated into "composition cells" and "leaf cells" which are, respectively, combinators and functions. Composition cells do not have any hardware associated with them; they simply provide a way to compose new cells from old ones through clever interconnection.

Although this work provides a formal mechanism for the discussion of hierarchy, it is primarily a tool for geometric design. In some sense, designs using composition cells are simply nested rectangles. This technique lacks the linguistic component present in a modern programming language which suggests the structure and semantics of a computation.

Hewlett Packard ASAP

The Hewlett Packard ASAP system resembles the Caltech Bristle Block approach [Bla81, Har80]. ASAP is a tool for cell level logic design which does not place any restrictions on the architectural layout of symbols (cells) as in Bristle Blocks. The designer manipulates symbols which are the same size and shape as the physical circuit cells they represent with their connection points indicated around the perimeter. By choosing a uniform width for symbols and standardizing the placement of connection points, interconnections are formed merely by placing the circuits adjacent to one another. This reduces local wiring (which must be routed manually in ASAP), but increases the complexity of the cell design task. The behavioral description of a symbol is restricted to the specification of which connection points are inputs or outputs and their fan-in or fan-out.

The Hewlett Packard ASAP system has been applied to 15 separate designs. Average design time was cut by 50 percent and component density increased as compared with handcrafted circuits. Most of this improvement was attributed to good local wire management.

Digital Equipment Corporation

The VLSI Advanced Development Group at Digital Equipment Corporation has produced a proprietary method and tools for the design of nMOS systems [Sta81, Mud81]. Although details about this method are sketchy, it appears that floor plans are developed in a top-down manner with careful attention to interconnection before starting the detailed layout. A chip assembler (based on composition cells) is used to generate the floor plan elements and circuit layout thus avoiding a great deal of time consuming hand layout.

Experience with this system indicates that layout at the floor plan level and the resolution of global wiring were essential to quick, successful design. The group recommends careful planning for power and clocks early in the design process. Hierarchical design tools and regular structures were identified as important contributions to designer productivity.

Xerox PARC

ICARUS (Integrated Circuit ARtwork Utility System) is an artwork editor for integrated circuit systems developed at the Xerox Palo Alto Research Center [Fai80]. Beginning with a hand-drawn stick diagram, the designer interacts with ICARUS and produces a detailed circuit layout. The design is presented to the engineer through two display windows. One window provides an overview of some large portion of the layout at a small magnification factor. The larger main window displays a smaller part of the layout with more detail. The designer can annotate the layout with text by labeling subsystems, signal paths, etc. The text is not interpreted in any way for analysis or interconnection.

ICARUS has been applied to the design of many VLSI systems at Xerox PARC and has been used to lay out several of the Multi-university Project

Chips (MPC). Unfortunately, ICARUS does not address rule checking, simulation, or the behavioral and structural characteristics of a VLSI system beyond its circuit geometry. Several other artwork systems which are similar in concept to ICARUS have been developed. These systems are CAESAR [Ous81, Ous82], Chipmonk [Pet82], and KIC2 [Kel82].

Princeton

Constraint-based layout is a new alternative to sticks and artwork editors. In this style, the position of an object (e.g., a box, transistor, gate, or subsystem) is described in terms of its relative placement with respect to its neighbors. These relations include "touches," "above," "below," "to the left of," and "to the right of." The designer is not permitted to specify an absolute size for an object nor an absolute position within the coordinate plane defined by the chip boundary.

A design tool for constraint-based layout has been developed at Princeton [Lip82]. Circuit descriptions are written in a relational placement language. The design tool, ALI, translates the description into the circuit geometry by reformulating the spatial relationships into a system of linear inequalities. Circuit primitives are translated to a set of cells that can be deformed and joined according to the solution of the geometric details about a design at the description level decreasing the amount of bookkeeping to be performed by the designer.

IBM

At IBM, work on logic synthesis has examined the successive transformation of register transfer level designs into "simpler" ones. Unfortunately, no specific results or algorithms have been reported [Dar80].

Research on time-domain circuit analysis is an attempt to reduce the amount of computing time required to get accurate estimates of electrical circuit behavior [San80]. By using nested macromodels, hierarchical system structure can be exploited to subdivide circuit analysis into smaller pieces which can be analyzed separately. Experimental implementation of the technique shows a significant savings in computer resources.

The late arrival of signals to a register is a problem in circuit design because if the data arrives late, then an incorrect result will be stored. IBM researchers have developed an algorithm to calculate the maximum delay through a combinatorial network between two or more registers, in order to verify the timely arrival of information at the destination registers [Rot81]. This algorithm combines the individual delays into an aggregate delay for the network.

MIT

Work at MIT has concentrated in several areas: simulation, artwork-based design tools, wire routing, and register transfer CAD tools. A logic level simulator for nMOS VLSI, called MOSSIM, was developed by Randy Bryant [Bry80a, Bry80b]. MOSSIM simulates a VLSI system at the MOS transistor level and uses three states (0, 1, and X) to represent signal levels. Simulation at this level provides more information about system performance than a gate-level simulation, but consumes fewer machine resources than continuous system electrical simulation.

Clark Baker and Chris Terman have developed a set of artwork-based tools that use the circuit geometry as the design database [Bak80a, Bak80b]. Their tools translate the geometry into an electrical network of nMOS transistors suitable for simulation on MOSSIM. During translation, several electrical and geometric analyses are performed such as checking for reasonable transistor size ratios and short circuit detection. This approach eliminates any conceptual inconsistencies between the geometry and its simulation descriptions. However, the tools consume an enormous amount of time deriving information about a design which could be more efficiently generated from a higher level description of the system.

MIT has produced some very encouraging results on wire routing. A linear time algorithm has been developed for channel routing and placement [Lei81]. This common situation arises when a set of pairwise connections must be made across a rectangular channel. In related work, algorithms have been devised for routing in rectilinear channels with less stringent topological constraints [Lap80, Pin81]. In the rectilinear case, an estimate for routing quality can be accurately calculated for a given placement without actually routing the wires. This is an important step toward fully automatic placement since it would help the placement tool to explore an entire space of component arrangements. The general placement and routing problem is NP-complete, however [Lap80].

The MIT Data Path Generator was constructed to support the development of the Scheme-81 LISP machine [Shr82a, Shr82b]. At the beginning of the design cycle, the engineers write the microcode for the machine using a LISP-like microprogramming language. In the second phase, the microcode compiler analyzes the microprogram and produces a description of what registers are used and how computations are performed upon the contents of those registers. In the third phase, the designers manually produce a description of the data path making space and time trade-offs. The Data Path Generator is applied to this description and performs the actual layout and interconnection of the cells within the data path circuit. Finally, the microcode compiler translates the microprogram to the bits required to control the registers and data operations.

A more refined version of this system is under development at the MIT Lincoln Labs, called "MacPitts." Using a dialect of LISP as the source description language, the MacPitts compiler translates a chip description to its data path and controller geometry without manual intervention [Sis82]. The data path is translated to circuit cells selected from a standard cell library and the control portion of the design is translated to a gate array. Locally, the circuits are interconnected using a channel routing tool. A more general routing tool is used to connect the major components with the input and output pads.

Users of these tools rely upon functional and electrical simulation for performance evaluation and debugging. It is difficult to relate the simulation results, which are based upon the primitive circuits generated by the compiler, to the symbolic description. The designer is forced to reverse the compilation process to effectively evaluate the simulation results. The linear, symbolic notation employed by these tools inhibits the incorporation of spatial (horizontal) and pipelined concurrency into a design because the description language is based upon a strictly sequential, register transfer model. Hence, if a given design is too slow, the designer may not be able to restructure the design and incorporate additional concurrency.

Stanford University

LAVA is a general purpose layout language for nMOS circuits [Mat82]. A circuit is described in terms of gates or stick diagrams with explicit interconnections. Acting upon a circuit description, the LAVA language processor performs several functions:

1. Extract the circuit from the gate and stick level description.
2. Check to see if the signal types match.
3. Compact the stick descriptions.
4. Stretch and abut cells.
5. Route the wiring.

Since LAVA is just a layout language, it does not directly address control. However, in a related effort at Stanford, Hennessy has developed a microcode compiler and simulation system called "SLIM" [Hen81]. SLIM uses a state machine description to produce a PLA-based microcontroller. The geometry of the controller can be included in a LAVA design thereby forming a complete digital system. Unfortunately, LAVA and SLIM have not yet been integrated through a common design notation to form a complete CAD system.

Carnegie-Mellon University

H.T. Kung has devised several techniques for the translation of signal processing algorithms into a very regular VLSI structure called a "Systolic Array." Each element in the array is a small processing unit which can communicate with its neighbors [Kun80, Kun81]. Although this work has produced some elegant translation rules and complexity results, it is not an effective tool for the design of systems with very little internal regularity such as a microcomputer or a communications processor. Application of this and similar techniques does hold promise for the design of special purpose signal processing devices [Cap81, Wei81].

C-MU is extending its RT/CAD system to VLSI. This work began with the thesis of Mario Barbacci in 1973 [Bar73, Bar74] on the translation of ISPS descriptions to a register transfer module set (DEC RTMs, also known as the PDP-16). In 1977, Edward Snow in his dissertation devised an internal representation for ISPS descriptions called the "value trace." The value trace is a data dependency graph which describes the nature and sequence of a digital computation [Sno78]. Subsequent work has examined the generation of data path and controller implementations from the value trace to MSI and LSI level circuits [Par79a, Par79b]. At present, RT/CAD can translate an ISPS description to a set of standard (gate level) cells. Translation tools which exploit the full generality of VLSI have not yet been reported in the literature.

ISPS and other register transfer notations have some disadvantages for VLSI design. A textual language is a weak notation for system structure. The linear text format hides all the structural information which can be provided at a glance by a block diagram. In a VLSI system where the interconnections play such an important role in layout and system performance, this structural information is crucial to the successful formulation of a design. Further, the text obscures the space and time trade-offs that are present in any VLSI design.

In another effort at C-MU, Sproull and Frank have constructed an integrated debugging environment for circuit testing [Fra81]. The designer first uses any of the standard geometry editors to develop a set of masks. The mask geometry is submitted to a circuit extractor which produces an electrical description of the device. The extracted circuit is simulated at the gate and logic level using FETS. This simulator creates a debugging environment for the VLSI designer with many of the features that one finds in a good programming system. The designer can interact with the simulation: examining and changing state, executing performance experiments, etc. Frank and Sproull intend to extend their tool to capture and incorporate simulation data into the functional testing and debugging of real circuits.

Summary

Digital system design is performed in three different decision domains: behavioral, structural, and physical. Due to the particularly heavy influence of physical decisions, tool builders have attempted to unify design in the physical and structural domains. The developers of the block concept have unified all three domains. Several researchers have settled on a symbolic logic approach in which the interior layout of a circuit cell has been eliminated, leaving connection points placed around a bounding box. The tools encourage hierarchical design through the use of nested symbols.

Unfortunately, these design techniques have put too much emphasis on layout and circuit matters. Some of the problems that remain are:

—Programming: One designer reports that 50 percent of development time was spent writing the microcode [Cla80].

—System control: Although the use of self-timed systems has been suggested in place of global clocks by Mead and Conway, none of the tools support locally synchronous, globally asynchronous control architectures.

—Higher level design styles: Unification of behavior and structure with physical considerations tends to force design at a fairly low level. In software, this is analogous to features of the host engine peeking through a higher level language.

To explore and solve these problems, the d technique and tools for the design of VLSI systems are proposed.

3

The d Method for VLSI Design

This chapter briefly discusses the origins of the method and its goals, and introduces the d method itself.

Motivation

In the integrated circuit design community there is the popular notion of a "silicon compiler" which would translate a high level specification for a chip into artwork. The specification would be written in a formal language roughly equivalent to a modern programming language such as Pascal. A compiler of this type would fully automate the IC design problem, significantly reducing designer effort and shrinking project schedules.

Unfortunately, a true silicon compiler does not yet exist. Present design systems, as language processors, are more at the level of an assembler—a translator that makes a one for one transformation from a simple description to its physical implementation [Rem79, Rem 81]. A higher level language and its compiler do many things for a programmer that assembly language and an assembler do not:

—Compilers perform bookkeeping tasks such as register allocation, memory layout, code improvement, etc.

—Control and data operations are raised from the machine representation level (e.g., branches and bits) to a conceptual, problem-solving level.

Although software engineers are not yet satisfied with the expressiveness or "power" of their languages, the compiler has freed the programmer from bookkeeping and has made time for the more stimulating task of problem solving.

VLSI design automation (DA) systems still put too many low level, trifling tasks on the designer. Commercial DA systems (especially) force the engineer to design a system strictly in terms of its geometry. It is difficult to view a complex of layers and boxes as an electrical network much less as a digital system. These DA systems do not possess a notation that delays the binding from the conceptual design to its physical, geometric form. Opportunities for analysis are also severely curtailed, since the geometric design is too low level to express the real, conceptual intent of the designer.

There are two other significant differences between existing VLSI design systems and the modern compiler. A programming language presents a self-consistent view of a program and its execution environment to the programmer [Dav80]. Language designers take great care to provide a set of data and control structures which have well-defined and nonconflicting interactions. If the designers are successful, the combination of language features in a program will result in greater "power" than the mere sum of its parts. Further, the translation from program to host machine instructions is total. All the information required to form an executing program is present in the higher level language program, compiler, and runtime environment. No missing pieces or intervening hands are needed to complete the translation. Existing VLSI DA tools do not always provide for microprogramming and other control problems and, therefore, are neither self-consistent nor total. This work is motivated by these deficiencies.

Goals

The d system intends to support the designer in several different ways:

—Provide a structured design method and notation which are self-consistent.

—Supply rules for translation that take a system description expressed in the design notation to its realization.

—Remove detailed, physical design considerations from behavioral and structural design.

—Give feedback to the designer about space utilization and timing delays.

—Assist the specification and translation processes with software tools.

These goals are sufficiently broad such that experience with d can be extended to either a complete, designer-oriented DA system or to the internals of a VLSI compiler. As the sophistication of automatic programming has grown since the 1950s, one can expect the sophistication of VLSI DA tools to grow, also.

Before discussing d, it is important to state what specific problems are not considered for solution:

—Automatic placement of system components and wire routing are not performed.

—Space and execution speed information is simply reported to the designer; the design space is not automatically explored. The designer is responsible for design evaluation and initiation of redesign.

—Simulation procedures (or data) are not produced, but the generation of this information is not precluded.

Each of these problems represents a significant, long term research task and is therefore omitted from the scope of this effort. However, these elements should be incorporated into any real CAD system based upon d, and on this basis, suggestions are made for their eventual integration with d.

The engineer must intervene in the translation process because interconnections are not made for the user by the design system. Therefore, d does not perform a total translation from a description to implementation and cannot be called a compiler. However, d makes several contributions toward a silicon compiler:

—Development of a self-consistent, higher level notation for nMOS VLSI which incorporates both data and control operations.

—Limited translation (without wires) of a description to artwork.

—Analytical techniques to derive space and delay information from the specification of a system—information which can be fed back into the design process.

—The ability to mechanically administer and manage a large design hierarchy.

Method

Within the context of d, the design process can be separated into three major stages: specification, analysis, and production. (See fig. 5.)

—The specification stage (fig. 6) is driven by the initial, informal specification of the system requirements.[1] The engineers partition the system into smaller, lower complexity units using d-n block diagrams and subsystem floor plans. Subsystems of very low or "implementable" complexity are subjects for detailed design. This task produces a description of the low level system structure, behavior, and layout in the form of function graphs (hardware subroutines) and floor plans.

Figure 5. d Design

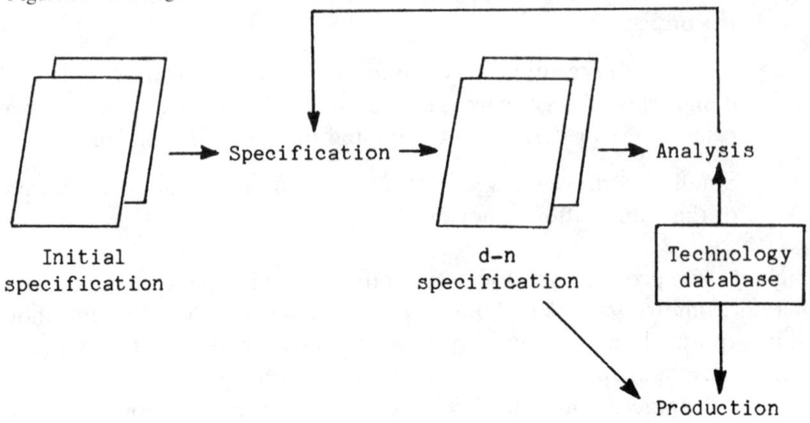

Specification
Analysis
Initial specification
d-n specification
Technology database
Production

Figure 6. d Specification Stage

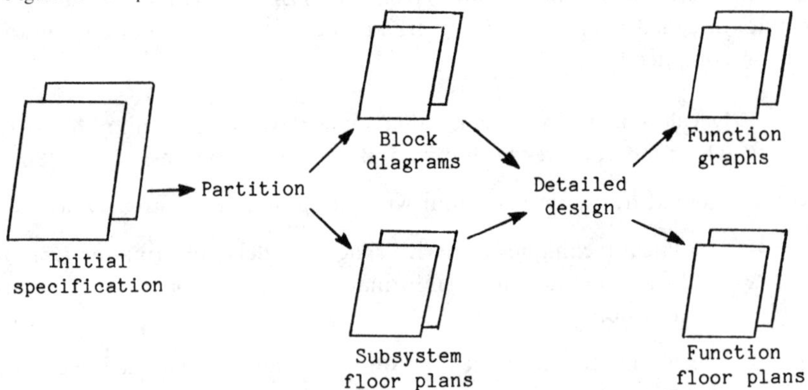

—The d-n descriptions, in conjunction with a catalog of primitive circuit elements called the technology database (d-tech-db), drive the analysis of the design. (See fig. 7.) The products of the analysis phase (e.g., delay analysis, functional and electrical simulation, design rule checks) are used to evaluate and modify the design.

—When the engineers are satisfied with the analytical results and the expected performance of their system, the d-n specifications and technology database are used to produce the circuit masks and prototype devices (fig. 8). After testing, the team may return to the specification stage to correct bugs or improve system performance still further.

Figure 7. d Analysis Stage

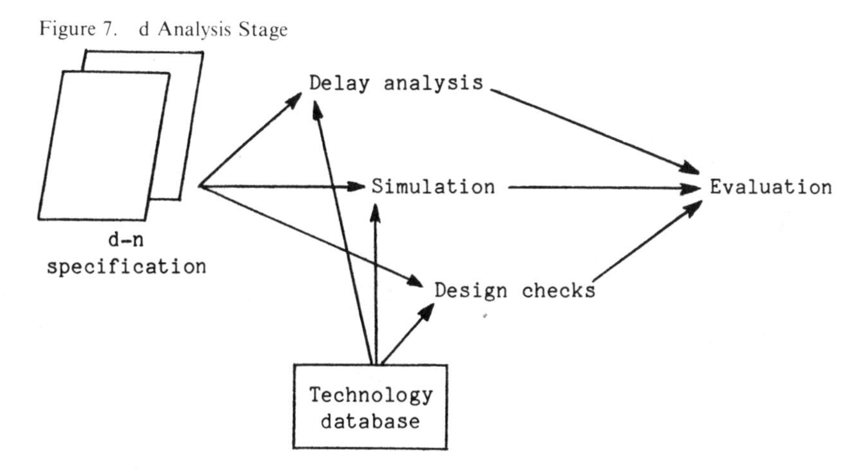

Figure 8. d Production Stage

d-n
specification
Mask
generation
Masks
Fabrication
Devices
Technology
database

Since the design is expressed in a machine readable form (d-n) and the technology database can be automated and shared, each step of the design process can be assisted by one or more software tools. This is a significant departure from current practice which employs many different descriptions and notations for design, analysis, and production.

d addresses another problem which is not adequately covered by contemporary design tools—the partitioning problem. Through d-n block diagrams, function graphs, and floor plans, system decomposition (top-down) or synthesis (bottom-up) can be mechanically assisted with a graphical editor. Interface protocols can be standardized and d-n designs can be checked for syntactic consistency at the subsystem interfaces. Further, d permits the estimation of execution delays from the d-n descriptions without extensive and expensive electrical simulation. This added capability should assist the system partitioning task where major space and time trade-offs are made.

This completes the general introduction to d. The following chapters describe d-n, the support tools, and an extended example.

4

The d-n Notation

This chapter presents a tutorial introduction to d-n, the notation for d-based VLSI design. (A more detailed description of d-n can be found in app. A.) The following sections discuss the style of design that d-n encourages, the description of system structure and behavior, and the translation of d-n descriptions into nMOS VLSI circuits.

Datatypes and Style

The descriptive style employed within d-n was inspired by the abstract datatypes of Guttag [Gut77]. With abstract datatypes, an object is characterized by the functions which may be applied to the object and by the axioms which govern the behavior of the functions. When designing with abstract datatypes, the system is decomposed and refined in terms of data objects and their functions. Eventually, a primitive level of refinement is achieved where the specifications can no longer be decomposed. Specifications at this level deal with primitive datatypes—objects that are native (fundamental) to the computing environment such as bits, integers, characters, or floating point numbers.

Unlike abstract datatypes which use axioms to describe an object in abstract, functional terms, a d-n datatype is concerned with the representation of an object and the computations performed on that object. Although the d-n description of a subsystem or function is operational in nature, it retains the style of abstract datatype specifications.[1] This style is used at every level of the design hierarchy from the interior of the hierarchy, through the leaves, to the primitive objects at the lowest level.

The Structural Hierarchy

When engineers approach a design problem for the first time, they tend to think of the overall system in terms of its subsystems and their interactions. These subsystems are further refined into additional subsystems until the design problem has been subdivided into manageable, low-complexity units.

In d-n, this structure of subsystems and communication paths is captured in a set of block diagrams called encapsulations.[2] Diagrams, rather than text, are used as the descriptive medium because:

—the structural design is visually apparent in a diagram,

—the complexity of the subsystem interconnections is reflected in the complexity of the diagram, and

—engineers are accustomed to block diagrams as a design aid [Est77].

The diagrams contain three kinds of symbols: subsystems (boxes), interfaces (arrow heads and tails), and connections (arcs). (See fig. 9 for a sample structural diagram.) The connection arcs, which can only be drawn between two interfaces, specify the interconnections between the subsystems and from the subsystems to the external interfaces. The direction of an arc, from its tail to its head, signifies a callee to caller relationship.

Every subsystem is an instance of a datatype, which must be described in its own d-n definition. d-n enforces hierarchical design through the notions of instance and definition in the following way. If definition A includes an instance of some other definition B, then A cannot be instantiated in B or within any of the subsystems of B. This property applies to those instances created directly by B or implicitly through one or more levels of subdefinition and instantiation.

In a style consistent with representational datatypes, the interaction between two subsystems, which is denoted by a connection arc, is modeled in the behavioral domain by a function call. There are, of course, two sides to such an interaction, namely, a caller and a receptor. Symbolically, a call is represented by an arrow head pointing into a subsystem, indicating that some measure of functionality is invoked by that subsystem. A receptor, which is an instance of a defined function that is specified at some lower level of the hierarchy, is represented by an arrow head pointing out from a subsystem instance. In this case, some functionality is being made available from the subsystem.[3] Physically, the arcs are translated to a communication bus leading from the function caller to its receptor.

The encapsulation diagram of a von Neumann-style computer is portrayed in figure 9. The heading at the top of the diagram states that this description is an encapsulation with the name "Computer." The encapsulation has three subsystems:

—"CPU," an instance of the encapsulation with the specification named "CPU-spec" (see fig. 10).

—"Memory," an instance of the defined datatype "Memory-spec."

—"IOchannel," an instance of the defined datatype "IO-spec."

Figure 9. Encapsulation Diagram for a von Neumann Computer

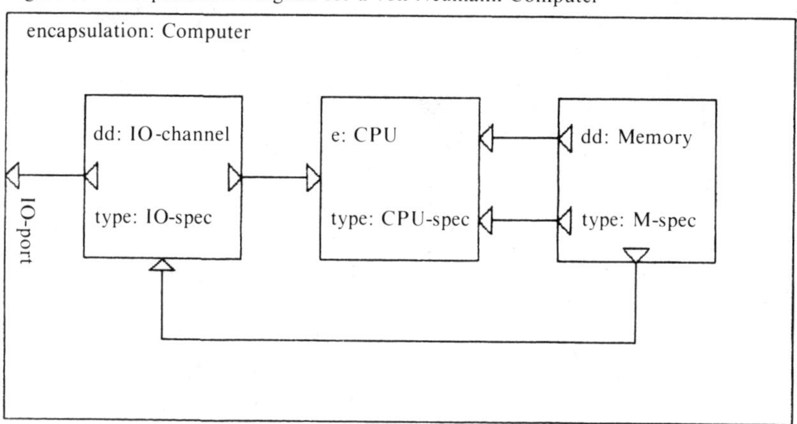

In the example, the function "IO-port" is made available from "Computer" for external use. Each connection between an interface of "CPU," "Memory," and "IOchannel" is represented symbolically by a connection arc.

The description of low level structure and system behavior is reserved for defined datatype or DD descriptions. They are the "leaves" of the design hierarchy. The decision to specify a subsystem as a defined datatype instead of an encapsulation is subjective. For each new subsystem (or type of subsystem) invented during system decomposition, the designer consciously or unconsciously asks the question:

> Is the behavior and the structure of the system function assigned to this module simple enough to be implemented directly in circuit cells?

If the answer to this question is "yes," the subsystem should be specified as a defined datatype. If the answer is "no," additional decomposition and refinement is required and an encapsulation should be used. During bottom-up design, this question is phrased differently:

> If this module was available in hardware, can I synthesize the higher level functions required by the design problem?

Figure 10. Encapsulation "CPU-spec"

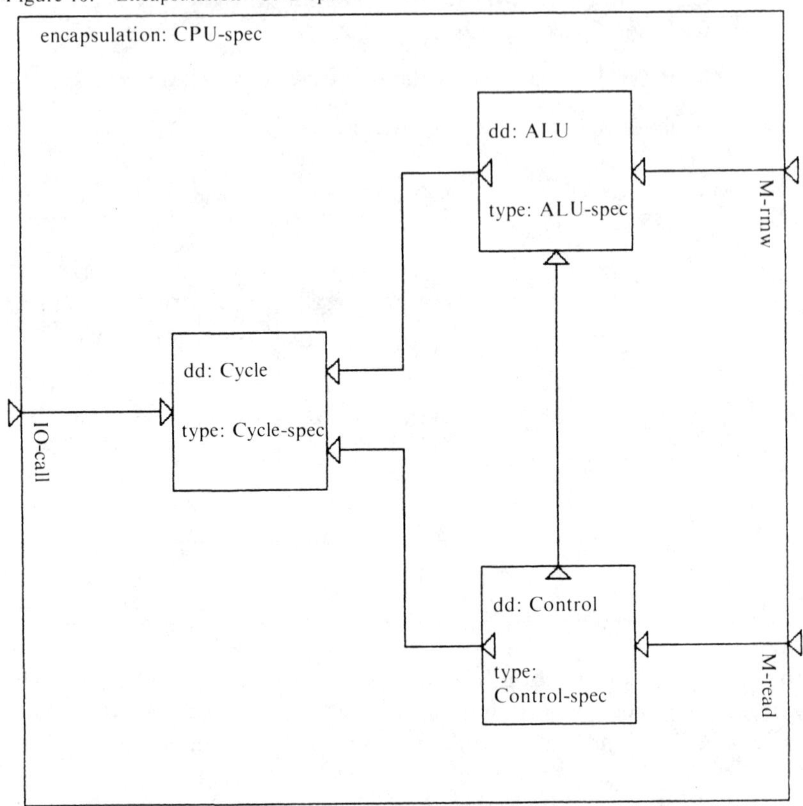

Again, if the answer is "yes," a defined datatype description should be used.

The defined datatype structural diagram is an extension of the encapsulation diagrams that describe the higher level structure of a VLSI system.[4] Defined datatypes are composed of one or more defined functions, storage and arbitration elements. The interconnections between these elements and the external interfaces are defined in the structural diagram using more interface points and connection arcs.

Figure 11 contains the structural diagram for the defined datatype "Control-spec." This DD implements the instruction fetch portion of the von Neumann fetch-execute cycle. It consists of two defined functions, "Fetch" and "ChangePC," a storage element, "PC," and an arbiter, "Mutex." There are five interface points placed around the box representing "Fetch" and five connection arcs extending from these points. The interfaces and connections have the following interpretations:

1. The external connection labeled with "Fetch" makes it available for use outside of the DD.
2. The external connection labeled with "Inst-read" represents a call to that external function. When it is called from "Fetch," it returns the instruction value from memory.
3. The internal connection "PC.Read" connects "Fetch" with the storage element "PC" through the arbiter. Execution of "PC.Read" gets the current value of "PC."
4. The internal connection "PC.Write" also connects "Fetch" with "PC" through the arbiter "Mutex." Execution of "PC.Write" within "Fetch" will write the new program counter value into "PC."
5. The external connection labeled with "Execute" is a call to that external function. When it is called, it will cause the instruction value to be decoded and interpreted by the arithmetic-logic unit.

The defined function "ChangePC" is called from outside of "Control." It is used to write the program counter register as a result of a jump or branch instruction.

This example illustrates several important characteristics of a d-n-defined datatype and the style of design that it encourages. Every storage element in a DD structural diagram is an instance of a primitive datatype and consists of:

—a storage object (circuit), and

—a set of primitive datatype functions that are used to manipulate the object.

The only way to write or read a storage element such as a register, for example, is through a primitive datatype function. d-n does not provide a predefined set of primitive datatypes like Pascal, Ada,or any other strongly typed programming language. It merely provides a graphical framework for working with primitive datatypes. The representation and semantics of a primitive datatype and its functions are left to the d-tech-db.

Access to an instance of a primitive datatype is strictly local and may only be manipulated through its primitive datatype functions by a defined function within the same DD. This constraint is enforced syntactically by the following rule:

> A connection arc may not link an interface belonging to a storage element (or an arbitrated storage element) with an interface point on the boundary of the defined datatype.

Figure 11. Structural Diagram for "Control-spec"

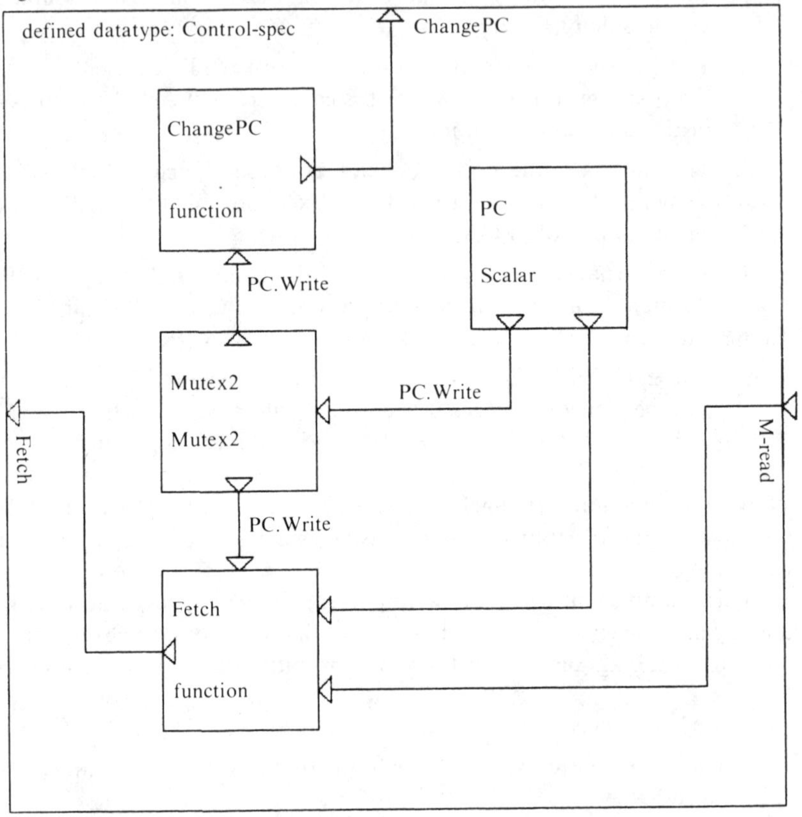

Through this rule, changes to system state are confined to the defined datatype level.

Since concurrent function activity is permitted, a mechanism must be provided to regulate access to a shared primitive datatype. Without arbitration, a race may occur when one or more defined functions attempt to use the information stored in a datatype while another function is changing that information. Symbolically, an arbiter has one interface point for every controlled access port and an interface to be connected to an instance of a primitive datatype. When a defined function executes a primitive datatype function node, it makes a request for its execution to the arbiter. The calling function suspends its operation until the primitive function has been executed and the arbiter returns a completion signal. When two or more primitive datatype function calls have been requested, the arbiter decides which request should be serviced first. The scheduling scheme employed depends upon the definition of the arbiter in the technology database.

The Leaves of the Hierarchy

There are two kinds of diagrams in a defined function specification: one each for the logical and physical description of the function. (The physical diagram will be discussed later.)

Figure 12 portrays the function "Fetch" from the defined datatype "Control-spec." Aside from the header, which assigns a name to the function and its diagram, the heart of the specification is the function graph. Input arguments are represented as labelled points and are arranged across the top of the diagram. In the example, the point labeled "Start" is the only input. The outputs, which are the result values produced by the function, are placed along the bottom of the diagram. "Fetch" has two outputs, "PC-value" and "Instruction."

Figure 12. Logical Specification of "Fetch"

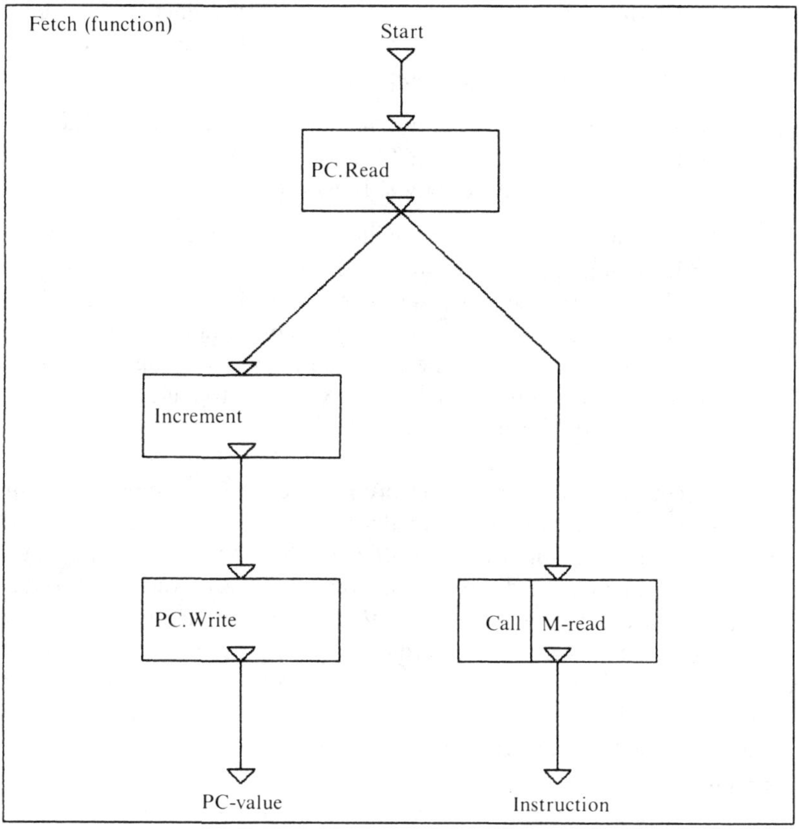

The rest of the function graph consists of a set of computational nodes and a set of interconnecting arcs. The nodes are drawn as boxes and the arcs are drawn as lines. Each of the nodes has a set of points arranged on its perimeter. These points are called ports and represent places where a data value is supplied to or removed from a node.[5] Input and output ports are placed on the top and bottom edge of the box, respectively.

The behavior of a node depends upon its definition in the technology database. As in the case of primitive datatypes, d-n does not have any predefined primitive nodes. It just provides the graphical, syntactic framework for using d-tech-db elements. There are three different kinds of nodes:

1. Primitive functions perform combinatorial data operations. Some examples of a primitive function are "Add," "Nand," "Increment," and data selection. They are strictly input-output functional and are not history sensitive—the result produced by a primitive function depends only on the input values. A primitive function does not have any memory elements which can recall some previous event in its history and perturb the computation.

2. Primitive datatype functions change or retrieve the contents of a storage element. All access to a storage instance must be funneled through its primitive datatype functions.

3. Operators invoke functions, conditional execution, and iteration. The rules regulating the kinds of permissible graph topologies prohibit cycles. Iterative behavior is impossible to specify if both graph cycles and recursion are forbidden. (Recursion is not allowed due to the difficulties this would impose on the translation of logical descriptions to their physical form.) The only recourse is to include an operator in the d-tech-db for iteration.

All nodes consume data values and produce new ones through their ports.

The arcs "conduct" these data values between the nodes and the function inputs and outputs. The arcs show the order in which the nodes must be executed to generate the desired result. Before a given node can produce its result, it must receive the data values that are its arguments. Any node that consumes the result must wait for the producing node to complete its execution. Therefore, a data producer must always execute before a data consumer. (Function inputs are producers and outputs are consumers by definition.) The function graph is a data dependency graph because it specifies the temporal and logical relationships between producers and consumers. The data dependencies provide much of the information needed to generate the circuit data paths and control state diagram for sequencing.

Figure 13 shows how a call to the function "Fetch" would appear in a function graph. It has one input port and two output ports corresponding to the input "Start" and the outputs "PC-value" and "Instruction" in the function specification. When the call node is executed, the value on the arc leading to the input port is passed to "Fetch." The producer "Start" sends this value to the consumer linked to it, the primitive datatype function call "PC.Read." The output of "PC.Read" is another data value which is sent to the primitive function "Increment" and the call node, "Mem-read."

Figure 13. Call to "Fetch"

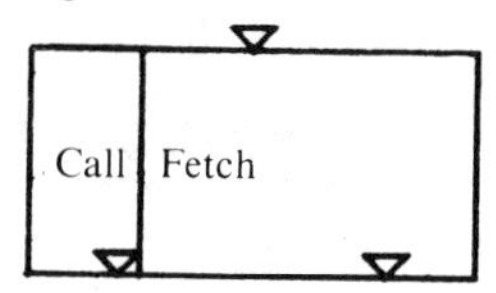

"Increment" adds one to the value it receives, forming the new program counter value to be written by "PC.Write." The call to "Mem-read" invokes a function in the memory unit that accepts a memory address as its argument and returns the word stored at that location. "Increment" and "PC.Write" can execute concurrently with "Mem-read" since a data dependency relationship does not exist between these two separate execution paths. Eventually, the two paths supply the outputs "PC-value" and "Instruction."

"Fetch" has one input and two outputs, but a function may have any number of arguments or results. A scheme must be devised to match ports at the site of the function call to the inputs and outputs in the specification. The following assignment rule is used:

> From left to right, assign each input (output) port at the call node to the corresponding input (output) in the function specification. If any nodes, inputs, or outputs are left unassigned, it is an error and either the call or the specification must be changed.

Hence, the order of arrangement for ports, inputs, and outputs is important since it governs the association of actual to formal function parameters.

Physical Design and Implementation

The physical description of an encapsulation, defined datatype, or defined function is a floor plan. In this kind of diagram, the actual dimensions and placement of the symbols are important. For example, the floor plan for "Computer" (fig. 14) is a scale drawing of the chip region which has been allocated for the encapsulation layout. The scale factor has been chosen such that the drawing fits and fills the page or display screen on which it is drawn.

Figure 14. Floor Plan for "Computer"

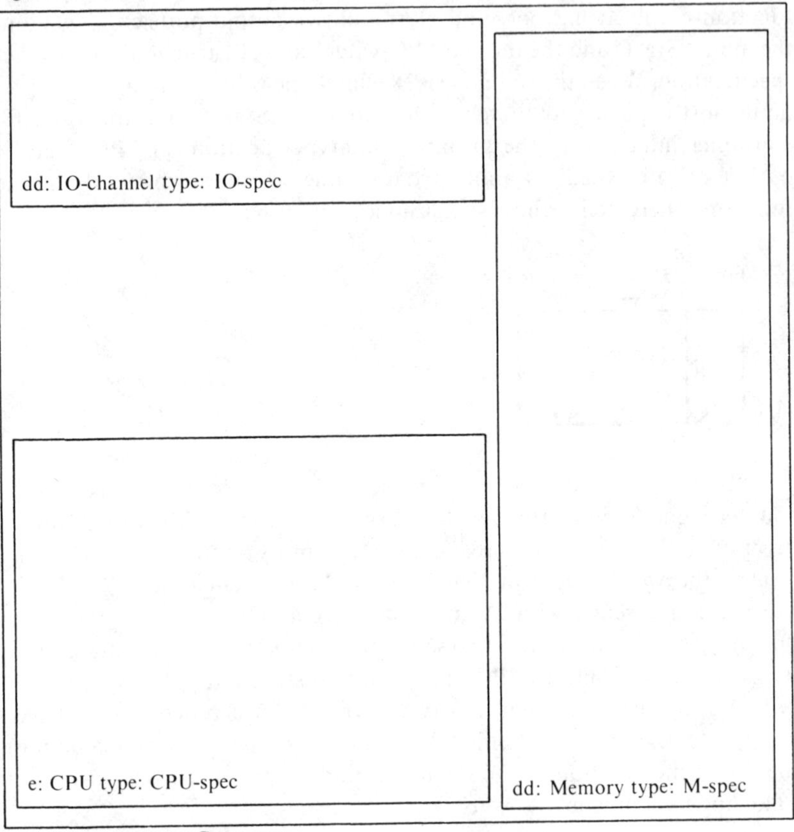

The design tools choose the appropriate scale factor for each drawing, but operate in absolute units (microns) internally.

The rectangles within the diagram boundary represent the subsystems within the encapsulation "Computer." The geometric origin of the diagram is the lower left corner of the boundary and all subsystems are placed relative to this origin. Subsystems may be rotated or reflected in order to make the layout more compact or to make wiring easier. All subsystems must lie completely within the boundary; no portion of a subsystem may exceed the region allocated to the enclosing encapsulation. This constraint permits incremental design rule checking [Whi81a, Whi81b, New82].

When physical design is first started, the size and shape of the region must be estimated because the exact sizes of the subsystems are not yet known. This estimate can always be adjusted as the size and shape of the subsystems are changed. Floor planning is the task of making estimates for subsystem geometry and then refining those estimates in response to changes in the detailed logical or physical design. For example, figure 15 contains the floor plan for "CPU-spec." If the size of "CPU-spec" is increased to accommodate a larger "ALU," the boundary of "Computer" may become too small and the instance "CPU" may exceed the boundary. "CPU-spec" must then be expanded, too.

Since d designs are hierarchical, space planning is also hierarchical. The encapsulation plan diagram is a natural medium for space budgeting. The details of low level subsystem layout are suppressed, letting the designer concentrate on the layout task one level at a time. The engineering effort applied to both the logical and physical design can be amortized across many copies of a subsystem because a single description can be used to create several instances of the same subsystem. Expansion of the hierarchy into a full "flat" floor plan is mechanical and can be performed by a relatively simple design tool. Full floor plans assist strategic layout decisions, possibly improving space utilization and simplifying the wiring between subsystems.

The physical description of a defined datatype is a floor plan, too (see fig. 16). It is a scale drawing showing the positions and sizes of the functions, arbiters, and storage elements in the DD. Although the size of a defined function must be estimated by the designer, the geometry for arbiters and

Figure 15. Floor Plan for "CPU-spec"

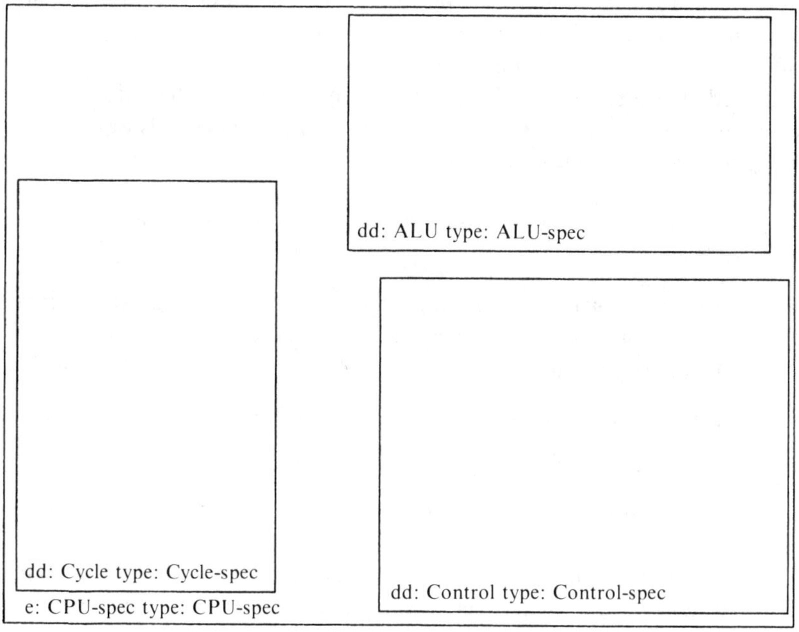

Figure 16. Floor Plan for "Control-spec"

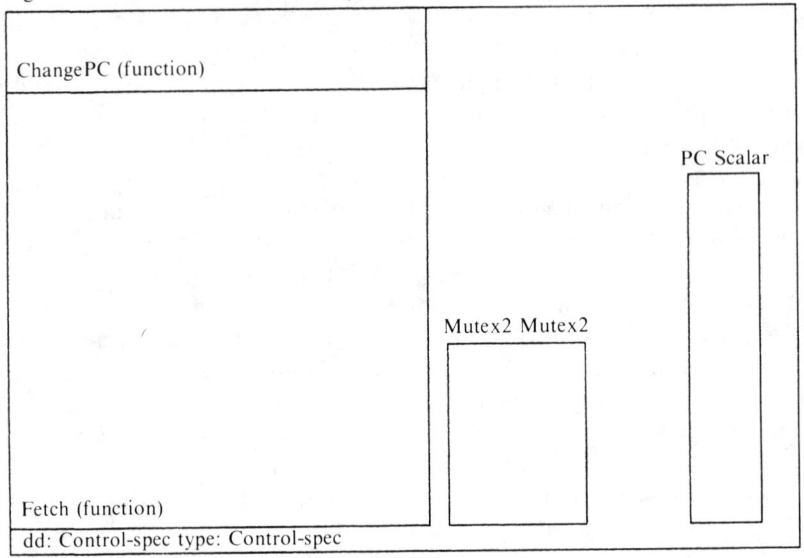

storage elements can be calculated precisely, because they are real, physical circuits. The arbiters, primitive datatypes, and datatype functions are described in the d-tech-db as procedural templates. (The primitive functions and operators are described in the same way.) In order to retrieve some information about a primitive object from the database, its template is executed. This process is called instantiation since the product of the procedure execution is an instance of the circuit. Several different kinds of products can be generated by a template, namely:

—Circuit geometry: A layer, box, and wire (LBW) description of the circuit.

—Space: Returns the maximum horizontal (X) and vertical (Y) extent of the circuit boundary. (Assumes that the lower left corner is the geometric origin.)

—Speed: Calculates and returns the worst case delay through the circuit.

—Control: Describes the control requirements of the circuit.

—Connection: Returns a list of connection points for wire list generation.

During instantiation, the designer or design system specifies the kind of product required and then executes that part of the template which produces it.

A template has a set of formal parameters whose actual values are specified by the designer at instantiation time. By parameterizing a template for gain, bus width, etc., a single cleverly written template can adapt to a variety of design situations. (The interested reader can find more about the procedural description of d-n primitives in app. B.)

For example, in order to find the physical size of "PC," the template for the primitive datatype "Scalar" must be executed. Figure 17 contains an excerpt from the d-tech-db template for "Scalar." The header declares three formal parameters:

1. "Command" selects the kind of product to be generated. In this case, it would select the routine labelled "Space."
2. "Name" is a string for annotating the instance.
3. "Width" is the number of bits to store.

Figure 17. Excerpt from the Template for "Scalar"

```
Template      Scalar
Object type   Primitive datatype
Parameters    Command, Name, Width.

Inputs        Fixed at 1. Data-in.
Outputs       Fixed at 1. Data-out.
Control       Fixed at 1. Enable.

Routine Space

      return < 90, 51 * Width >
```

When "Scalar" is instantiated with the actual value of the parameter "Command" equal to "Space," it will return a pair of integers which are the maximum horizontal and vertical extent of the instance. (The unit of measurement employed here is one-half of the minimum feature size, "lambda.") Note that the vertical extent depends upon the value of the parameter "Width" since the overall size of "Scalar" depends upon the number of bits that it can store. The coordinate pair returned by the routine can be used to draw a scaled rectangle in the defined datatype floor plan to represent the instantiated form of the storage element.

The physical layout of a defined function is captured in a floor plan diagram like the defined datatypes and encapsulation. Since the defined function is the lowest level of the design hierarchy, all parts of a defined function must be translatable to hardware. The translation of a defined function to its implementation involves three interrelated elements: instantiated templates (circuit cells), control, and wiring (data and control paths plus power distribution). The function floor plan shows the placement of its circuit cells, controller, and wiring channels.

Through the technology database, primitive function nodes, function inputs, and function outputs are translated to instantiated circuit cells. Operators and primitive datatype function nodes, however, are translated to control signals emanating from the controller since they do not perform any intrinsic data operations by themselves.[6] The function input and output cells are buffer registers that hold incoming and outgoing data values. For each cell, the designer must specify its actual instantiation parameters. In the case of the input and output cells, for example, the designer must specify the number of bits that are required to represent each data value.

Figure 18 shows the floor plan for the defined function "Fetch." Five cells have been instantiated:

1. The input "Start" is a one-bit "Entry" cell.
2. The primitive function "Increment" has been instantiated and is 8 bits wide.
3. The output "PC-value" is an 8-bit wide "Exit" cell.
4. The output "Instruction" is an 8-bit wide "Exit" cell, also.
5. The controller cell has been instantiated and programmed to sequence "Fetch" through its execution states.

The open space between the cells will provide channels for wire routing.[7]

The graph arcs define both the data dependencies and datapaths within the function. From the data dependencies and the control requirements of the nodes, the controller implementation is formed. The controller consists of a finite state machine (FSM) and a local two-phase, nonoverlapping clock. The clock period and state transitions must be chosen such that data values are produced before they are consumed. Due to the semantics of arbitrated primitive datatype function calls and defined function calls, the controller must be able to assert a function call request and then suspend its execution until a restart signal arrives sometime later. This style of controller is compatible with the one-hot scheme of Carter and Hollaar, whose tools can be used to create the actual controller implementation from the state table description of the FSM [HolIE, Car81].

Figure 19 contains the state table for the "Fetch" FSM controller. (The derivation of this state table and the graph to state table translation algorithm are presented in app. C.) The states Initial, Entry, and Exit are present in every

Figure 18. Physical Plan for "Fetch"

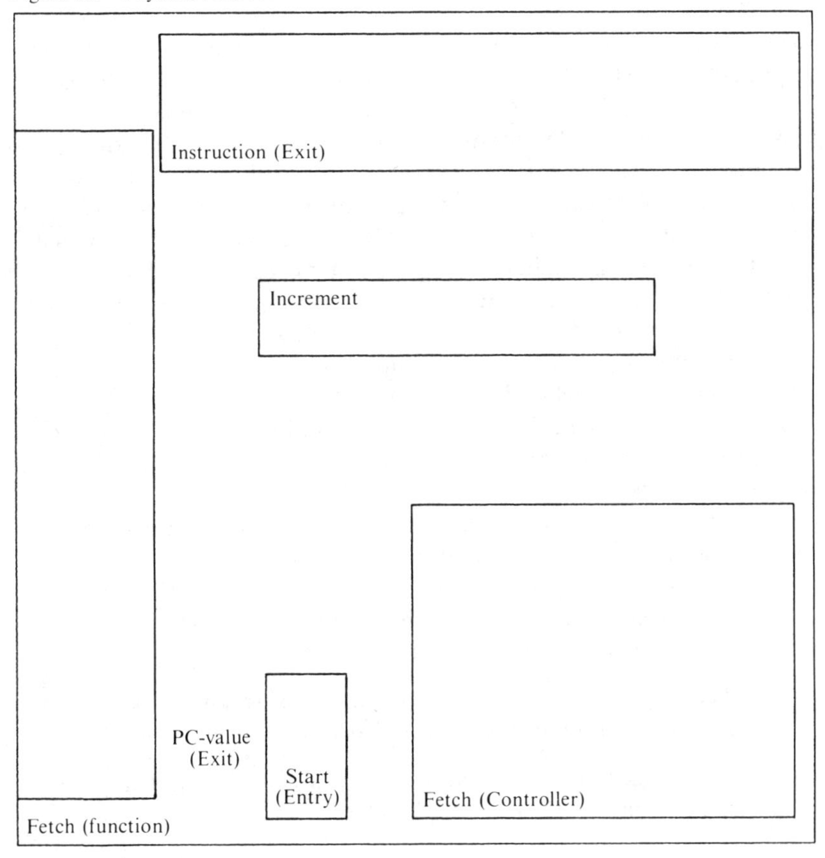

Figure 19. State Table for "Fetch"

```
States:

    Initial, Entry, Exit, 1, 2, 3

Transitions:

    Initial -> Entry : Call
    Exit -> Initial
    1 -> Exit : Synch 1 2
    2 -> Exit : Synch 1 2 Ack
    3 -> 1
    3 -> 2
    Entry -> 3
```

controller and are part of the function call signaling protocol. Initial is the idling state where the controller waits until the signal Call is asserted. It then proceeds to the Entry state which enables the incoming data values into the input buffer registers. From Entry, the controller goes to state 3 (PC.Read) and then enters states 2 and 1 (PC.Write and M-read) in parallel. State 2 has an acknowledgment condition (Ack) associated with it because the controller must wait in that state until the defined function "M-read" sends a completion signal. Next, Exit cannot be entered until both 1 and 2 are active as the synchronization condition "Synch 1 2" indicates. During Exit the outgoing result values are enabled into the output buffer registers. The Exit signal is sent to the function calling "Fetch" as its completion signal. The caller will subsequently drop Call, completing the 4-cycle signaling protocol.

The controller proceeds from state to state in synchrony with the local clock. Where a synchronizing condition (Ack or Synch) appears, that transition will not be performed until the conditions have been asserted. Since interfunction communications are asynchronous and intrafunction operations are synchronous with their own local clock, d-n systems are self-timed.

Using the internode datapaths (arcs) and the physical placement of the nodes, a local wiring list can be generated because:

—Each cell has a set of electrical connection points in a one to one correspondence with its input and output ports.

—The position of each connection point is known relative to the geometric origin of the defined function floor plan.

—The connection points can be physically interconnected according to the arcs in the function graph.

Similarly, the physical sources and sinks of the control signals, power, and ground can be located and interconnected.

Connections which are made outside the graph due to a function call or primitive datatype function must be resolved through the defined datatype structural diagram in the following way:

> Find the interface in the structural diagram that corresponds to the extra-graph connections at the site of the function call. Follow the connection in the structural diagram to the other interface. Determine the physical placement of the connections at that point. Draw wires from the connection points at either end, matching sources with sinks, bit by bit. Use the left to right ordering of function arguments and results to perform source and sink matching.

Local wiring is a part of a defined datatype like its functions, arbiters, and storage. Instantiation of a DD brings along a copy of its local wiring as well as a copy of its subcomponents. Because local wiring can be replicated, the engineering cost of wiring a defined datatype can be amortized across several instances of the DD.

Since a defined function may be called externally from some other defined datatype, a global wiring list must be created to connect the function with its caller. Once a defined datatype has been instantiated within the logical and physical design hierarchy, its structural and physical relationships with its neighbors become known:

—Its external interfaces are connected with other subsystem instances and their interfaces by connection arcs.

—The placement of the defined datatype becomes fixed with respect to some absolute origin such as the origin of the chip.

By unwinding the structural interconnections, a particular function instance can be associated with its caller. If the physical hierarchy is expanded at the same time, the electrical signal sources and sinks at the function call can be associated with their counterparts at the target instance and a wire list can be produced. (An algorithm to generate the global wire list is given in app. D.)

5

Software Tools

In the last chapter, the d-n design notation was discussed. A thread was established from the logical and floor plan description of a VLSI system in d-n to its physical implementation. This chapter proposes a set of tools to help the design team to specify a system, to translate its description to its implementation, and to evaluate the result against the system requirements.

As discussed at length in chapter 2, a VLSI system can be abstracted in several different ways. It is at once a physical device, a very tiny and orderly arrangement of semiconductor materials and a digital system. To make the VLSI design task somewhat easier, tools have been devised by the academic and industrial design community to help their designers to make decisions in each of these abstraction domains. For example, artwork editors and wire routers have been built to create the system geometry. Simulators have been developed to evaluate performance at both the electrical and register transfer levels.

Unfortunately, these tools were constructed by independent software development groups. Often, a tool is too specific to a particular abstract view of the system to be useful in another problem domain. For example, the description of a system as an electrical network is insufficient to generate the mask geometry. Within any given VLSI design project, it is not uncommon to find several different and possibly inconsistent descriptions of a system. The circuit extraction tools of Baker are an attempt to establish consistency between the geometry and its electrical description [Bak80a, Bak80b]. This approach cannot be extended, however, beyond the transistor level and, therefore, does not assist higher level design or analysis.

By using a standard notation (d-n) to represent a design, the d system departs from contemporary practice. Since the proposed d tools operate on a common system specification and technology database, consistency throughout the design process is assured. d-n also serves as a communication medium between designers and designs are more easily shared among the members of the project team.

A common design representation has still another advantage—the tools can be integrated together to operate as a self-consistent unit. Software engineers have been working on the concept of a workbench or programming system for software development. Some examples of a programming system are Interlisp, the Programmer's Workbench, and APL [San78, Dol78, Pak72]. These tools create an environment for program creation, debugging, and execution using intimate knowledge of the programming language that they support.

The proposed tools form a self-consistent engineering sketchpad for VLSI design. Within this environment, the designer should be able to specify, simulate, analyze, and evaluate a design. In every instance, the tools could refer the user back to the d-n description—the reference point for all design tasks. For example, when delay analysis is applied to the d-n specification, the results could be reported in terms of the system structure. The designer can then isolate potential bottlenecks in the design and correct the logical and physical definitions accordingly.

The power of an integrated design environment is in its ability to handle the conceptual transitions from one kind of abstraction to another. To help the designer make these transitions d should present several views of the system to the designer simultaneously through several different display windows. Inconsistencies between the windows could be flagged. The description editor, for example, would display the logical and physical parts of a subsystem specification in two separate windows. If a change is made in the logical specification that is not reflected in the physical one, the offending symbol could be displayed in a different color or could be made to blink.

The rest of this chapter proposes the d software tools from the perspective of the designer. An experimental set of tools was constructed to evaluate the feasibility of this new approach to VLSI design and to learn about its human interface issues. (See chap. 7.) The following discussion will draw from this implementation experience and the experience of other researchers whose tools are being applied in practice.

The logical and physical description of the system would be created and modified with a graphical editor, d-editor. With this tool, the designer should be free to move through the design hierarchy. The engineer should be able to bring up the specification of more than one encapsulation, defined datatype, or function onto the screen, because it is important to see the context of a subsystem while working on its definition in order to make decisions about structure or component placement.

A subsystem definition could be displayed to the user in two different graphics windows (fig. 20). The left window might contain the logical specification of a subsystem or function and the right window could display its floor plan. To avoid operator confusion, only one display window (of possibly

Figure 20. d-Editor Windows

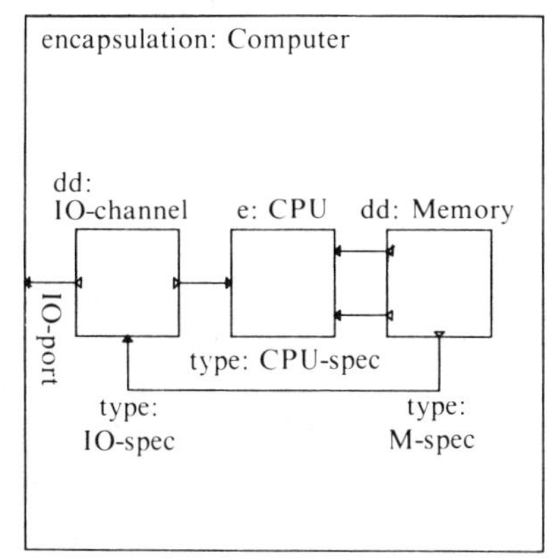

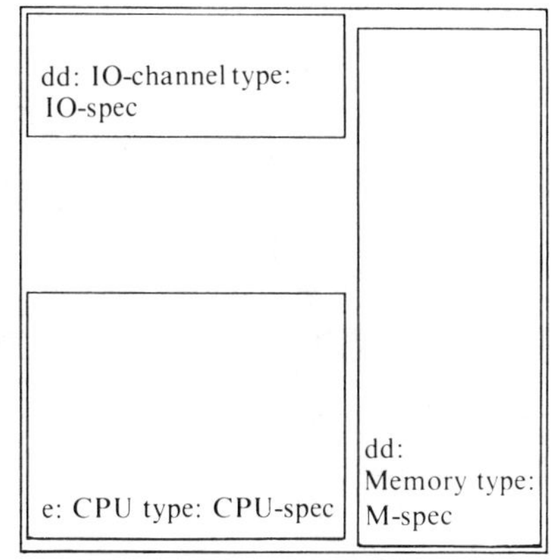

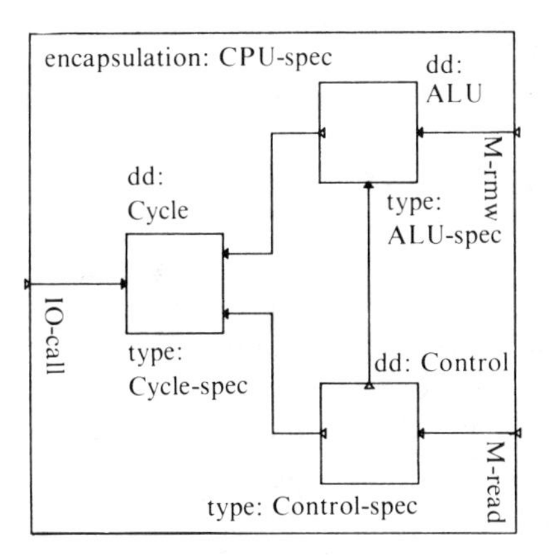

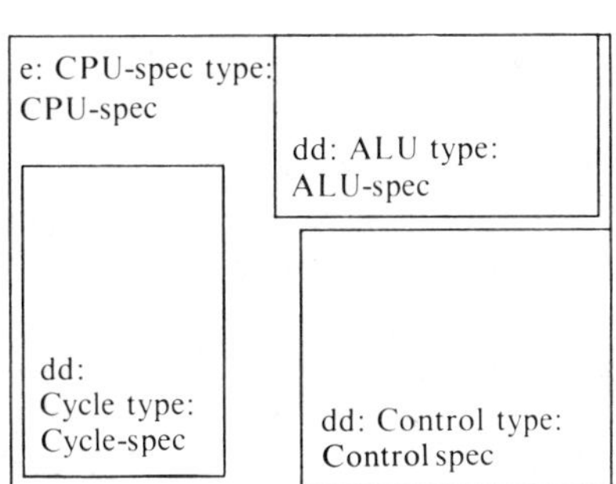

several windows and different descriptions) should be available for editing at any particular moment and may be selected by moving the display cursor within its boundaries.

The multiple window scheme has several advantages. The definition of an object, the context of its use, and the descriptions of its components could be reviewed by the designer together. This would help the user during decomposition or synthesis when many movements up and down through the design hierarchy are required. Experience with an experimental editor indicates that rapid movement from one subsystem to another is essential when defining interfaces. By displaying the logical and physical parts of an object description next to one another, the translation from the logical diagram to its physical layout would be visually apparent. Further, the placement and interconnection of design components in the logical diagram could be used to devise a floor plan which reduces the number of physical wire crossovers. Since the spatial context of a subsystem (i.e., the floor plan in which it appears) could be displayed at the same time, the subsystem plan could be adjusted to facilitate the global wiring, also.

The d-editor should detect and prevent many kinds of graphical syntax errors. In particular, illegal interconnections should be found such as an input connected to an input, more than one connection arc connected to an interface point, etc. Some kinds of errors are due to inconsistencies between the definition of a subsystem and its use. For example, a defined datatype interface which makes a function available for use by another DD must be labeled with the name of the function that defined the behavior at that interface. If a connection is made to this function in an upper level of the hierarchy, the interface in the enclosing encapsulation must be labeled with the same function name. To catch interlevel errors of this kind, a consistency checker (d-checker) should be included in the d package.

Although d-checker would detect and report the presence of overlapping subsystems in the floor plan, some design rule violations may still occur. For example, if manual wire routing is required, the designer could easily place some wires too close together or too close to a circuit cell. For this reason, a design rule checker, d-DRC, should be included in the d package. Due to the enormous volume of geometric data to be processed (500,000 rectangles), design rule checking is a resource intensive task. Regularity and hierarchy could be exploited to reduce the computational workload [Whi81a, Whi81b].

d should provide for downward extensibility by permitting the designer to create new circuit primitives for the technology database. This is an important capability because no one can anticipate all the future applications of the CAD system and construct a "complete" suite of cells. For example, new circuits are continually introduced into the standard logic sets (e.g., 74LS00 series TTL, ECL 10K, etc.) as the IC manufacturers try to satisfy the expanding needs of their customers.

A cell editing tool, d-cell, should be provided to accommodate the design of new d-tech-db primitives. Unlike other cell (symbol) editors which create a static description of the circuit geometry, d-cell would create the procedural descriptions used in the technology database. (Appendix B discusses the procedural cell model.) Since each cell entry supports several different database views other than its geometry, d-cell must help the designer to draw the floor plan view of the cell and to write its simulation descriptions, wire list routines, and delay formulas as well.

Eventually, a controller must be generated for each defined function in the system. The tool d-FSM would produce a state table for the controller and its equivalent nMOS implementation. The resulting state machine would then be placed into the floor plan for the defined datatype. The state table should be retained for simulation and delay analysis. (See app. C.) An experimental version of d-FSM was constructed that produces a control state table from a function graph. The state tables in this study were produced with the experimental d-FSM.

Two wiring tools would be provided by d. d-router would be an automatic routing tool that generates the wire lists for a design and then attempts to place wires between the connection points specified by the lists (see app. D). Until the automatic placement and routing problem is satisfactorily solved, d-router would use the floor plan created by the designer. (It has been shown to be an NP-complete problem.)

Routers have been constructed for printed circuit boards and gate arrays where the component placement is fixed by the designer (PC boards) or by constraints imposed by the technology (gate arrays.) Often, these routers are not sophisticated enough to find a complete set of wires for a given arrangement of components and connection points. Approximately 85 to 90 percent of a gate array can be successfully wired without human intervention [ShiIP]. Anticipating this need, a manual wire editor, d-wire-editor, should be provided in the d package. When d-router finishes a routing task, the failures could be reported to the designer who would use the d-wire-editor to complete the routing task. Alternatively, the designer could have the option to rearrange the floor plan with d-editor based on the failure report of d-router. Of course, once the floor plan has been changed, the original wiring lists will become invalid and the router must be run again.

The wiring tools could operate at three different levels: defined function, defined datatype, and global (inter-defined datatype.) Wiring changes due to layout modifications would, therefore, be isolated to a particular subsystem or function. A change to a defined function would not necessarily invalidate all the other defined datatype and global wiring if the position and arrangement of function inputs and outputs were kept intact.

Simulation and design analysis are especially important tools and should be included in the d package. For example, the electrical behavior of an nMOS circuit is extremely complex and does not readily admit to closed form, mathematical solution. Validation of the design cannot be performed on the actual circuits after fabrication because:

- The "design, fabricate, and test" loop is much longer than the equivalent software development loop and is more expensive.
- Some of the devices may not work due to fabrication errors or wafer flaws. The designer cannot always tell if the design is faulty or if the design is good and the chip as manufactured is defective.

For systems of a practical size (20,000 gates), simulation is a necessity.

The d CAD system should support both electrical and higher level, functional simulation. d-functional would simulate the system in terms of its logical behavior ignoring any of its electrical characteristics other than signal delay. Using d-functional, conceptual design errors could be found before time is spent on detailed layout, wiring, and electrical simulation. d-circuit would simulate the system (as specified in d-n) as an electrical network of transistors and wires:

1. If the wiring and layout have not been defined, the network structure could be derived from the logical part of the d-n description. Nominal (default) values would be used for path resistance and capacitance.
2. If the layout has been specified, but not the wiring, path lengths could be estimated as the Manhattan distance from one circuit cell to another. These estimates would supply the node capacitances and resistances.
3. If the layout and wiring have been defined, the actual path resistances and capacitances could be included in the network description.

Using the first form, the designer would be able to estimate the system performance without investing design time in detailed layout and wiring. Obvious design flaws would be visible at this stage. After a satisfactory first level result had been obtained or after the floor plans had been drawn the second level of electrical simulation could be applied to the design. Since more realistic capacitances and resistances would be included in the network, a better estimate of system performance could be derived without performing the relatively expensive wiring task. At the third level, a complete electrical description of the system could be generated resulting in the best possible simulation data.

Designers often calculate the expected delay through time critical functions before simulation. If the time constraints on those functions are not met, some portion of the system must be redesigned. Frequently a system is designed around its critical functions; satisfactory performance is guaranteed first, then the ancillary functions are added.

The d-n representation and the d-tech-db technology database provide the opportunity to mechanically calculate and report the expected execution speed of selected system functions. The delay analysis algorithm works in the following way.

1. The designer selects a particular function F for analysis.

2. If the clock period has not been determined for F, then find the longest combinatorial delay in its function graph. The clock period is twice the maximum delay time.

3. If F does not call any defined functions, its execution time is the product of the clock period and the number of transitions needed to execute F. Find the longest execution path through the state table for F from the initial state, through the transitions and intermediate states, and eventually returning to the initial state. The number of clock periods needed to execute F is equal to the number of transitions on the path.

4. If one or more defined functions (G_i) are called by F, calculate the execution speed for each of the G_i through the recursive application of this algorithm. For every path through F, calculate its aggregate delay by summing the transition and function call delays encountered along the path. The maximum aggregate delay is the execution time for F.

Although this algorithm is recursive it eventually terminates because the hierarchical design is free of cycles and recursively defined functions are prohibited. (App. E discusses the delay analysis algorithm in detail.) A delay analysis tool, d-da, would be included in the d CAD system. An experimental version of d-da was used to prepare the delay analysis of the cache memory system in chapter 6.

The following discussion illustrates the operation of the delay analysis algorithm as applied to the defined function "Fetch." In figure 21, the function graph of "Fetch" has been rewritten, replacing the inputs and outputs with the special nodes "Entry" and "Exit." (This rewritten form is used by the control generation algorithm to find the data dependencies in the graph. See app. C.) Assume that the designer is interested in the execution speed of "Fetch" and apply the delay algorithm to its specification.

First, the clock period must be determined. Using the template "Speed" routines in the technology database (figs. 22, 23), each of the primitive nodes is labelled with its combinatorial delay. The longest combinatorial delay between two state bearing nodes is:

$$45 + 128 = 173 \text{ nanoseconds}$$

due to the delays of the "PC.Read" and "Increment" nodes. Therefore, the clock period for "Fetch" is 346 nanoseconds.

"Fetch" calls the defined function "Mem-read." For brevity, assume that the algorithm has been applied to "Mem-read" and its delay has been evaluated at 500 nanoseconds. Two paths through "Fetch" are executable: one path to

Figure 21. Defined Function "Fetch"

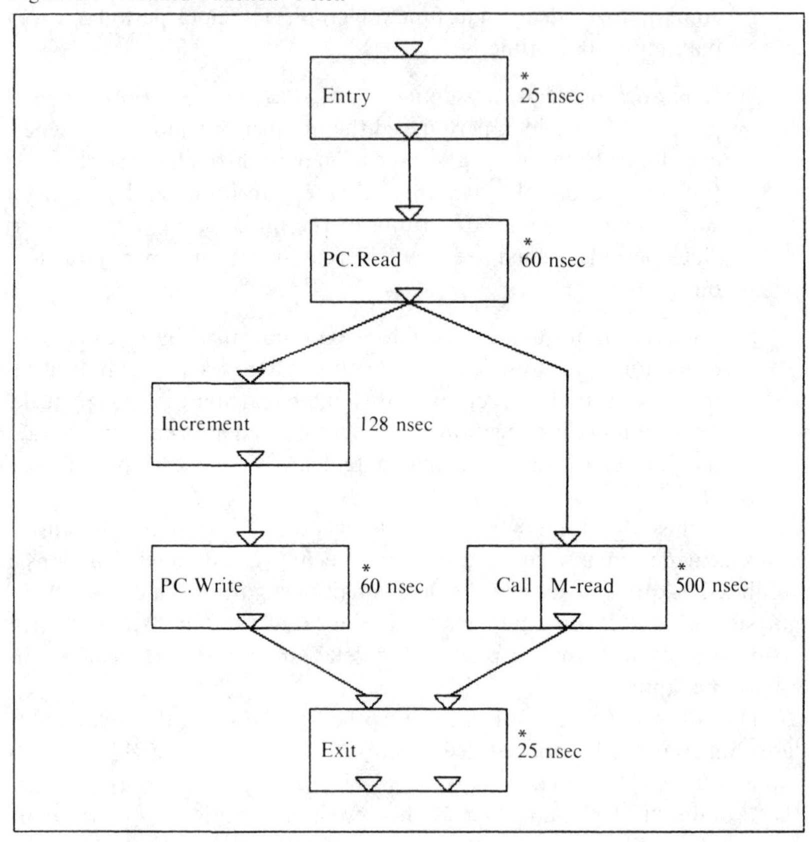

Figure 22. "Entry," "Exit," "Call," and "Increment"

```
Template      Entry
Object type   Special
Parameters    Command, Name, Width.

Inputs        Fixed at 1. Argument-in.
Outputs       Fixed at 1. Argument-out.
Control       Fixed at 1. Enable.

Routine Speed

    return ( 25 nsec )
```

Fig. 22 (contd.)

```
Template      Exit
Object type   Special
Parameters    Command, Name, Width.

Inputs        Fixed at 1. Result-in.
Outputs       Fixed at 1. Result-out.
Control       Fixed at 1. Enable.

Routine Speed

    return ( 25 nsec )

Template    Call
Object type Operator
Parameters  Command, Name.

Inputs      Variable.
Outputs     Variable.
Control     Fixed at 1. Enable.

Routine Speed

    return ( 20 nsec )

Template    Increment
Object type Primitive function
Parameters  Command, Name, Width.

Inputs      Fixed at 1. Integer.
Outputs     Fixed at 1. IntegerPlus1.

Routine Speed

    return ( Width * 8 nsec )
```

call "Mem-read" and one path to increment the program counter. These paths have aggregate delays of 2230 nanoseconds and 2076 nanoseconds, respectively:

2230 nsec = (346 nsec * 5 transitions) + 500 nsec
2076 nsec = (346 nsec * 6 transitions)

Figure 23. Delay Routines for "Read" and "Write"

```
Template      Read
Object type   Primitive datatype function
Datatype      Scalar
Parameters    Command, Name, Width.

Inputs        Fixed at 1. Request.
Outputs       Fixed at 1. DataValue.
Control       Fixed at 1. Enable.

Routine Speed

    return ( 45 nsec )

Template     Write
Object type  Primitive datatype function
Datatype     Scalar

Parameters   Command, Name, Width.
Inputs       Fixed at 1. DataValue-In.
Outputs      Fixed at 1. DataValue-Out.
Control      Fixed at 1. Enable.

Routine Speed

    return ( 60 nsec )
```

The path through "Mem-read" takes the longest to execute. Hence, the execution speed of "Fetch" is reported to the designer as 2230 nanoseconds.

With d-da the designer would be able to evaluate the performance implications of a particular logical design without resorting to simulation. The delay analysis algorithm could be extended to incorporate the effects of wire capacitance, since the combinatorial delay of a circuit primitive is computed by a template procedure and it can be parameterized with an estimate of the capacitive load that the circuit must drive. The cache memory example (chap. 6) shows how information about path material and length can be used to obtain a better estimate of system performance in this way.

Once the project team has become satisfied with the design and is ready to begin fabrication, the d-n description must be translated to pattern generation tapes. The software tool d-artwork would perform this final step.

6

Example: Cache Memory Design

In this chapter, d is applied to the design of a cache memory chip. Although the cache is not really complex enough to be regarded as a VLSI part, it is sufficiently large to demonstrate and test the features of d-n. To assess the feasibility of a real CAD system which uses d as its basis, some experimental tools were constructed.

In this chapter, the cache memory design problem is first described in English prose. This section will act as an initial specification for the example system. The second section presents the overall structure of the design—the structural hierarchy. Next, the defined functions at the leaves of the hierarchy are described. The implementation of the cache is discussed in the fourth section, followed by a sample delay analysis of the cache functions. In the concluding section, some comments are made on the evolution and implementation of the design.

Cache Memory Requirements

The purpose of the cache memory is to improve access speed between a processing unit, "Pc," and a very slow memory, "M." The memory is 8 bits wide and has a 16 bit address. Hence, all data and address busses wil be 8 and 16 bits wide, respectively.

As shown in the Processor-Memory-Switch diagram of the cache (fig. 24), there are three important interfaces. The data and control bus between Pc and the cache memory communicates read and write requests from the processing unit to the cache. The data and control bus between the cache and M communicates read and write requests from the cache to the memory. The address bus is shared by all three units. The processing unit sends the desired memory address on this bus to the cache chip and M. All inter-unit communications use a four-phase request and acknowledge protocol.

The cache memory is organized into four blocks (fig. 25). Each block has an 8-bit data field, a 16-bit address field, and a validity bit.

Figure 24. Cache Memory

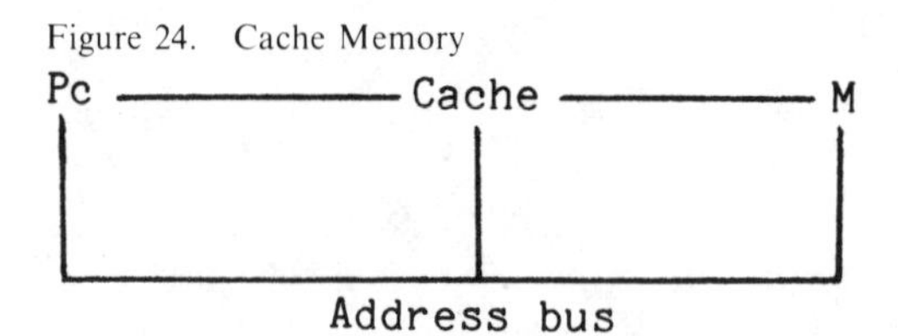

Figure 25. Cache Organization

0	$Address_0$	V_0	$Data_0$
1	$Address_1$	V_1	$Data_1$
2	$Address_2$	V_2	$Data_2$
3	$Address_3$	V_3	$Data_3$

To read from memory, the processing unit puts an address onto the address bus and requests a read operation. For each block with valid address and data fields, the requested address is compared with the block address. If a match is found, the value of the data field is transferred to the data bus between the cache and Pc. If none of the addresses match, the value is read from the memory unit and is sent to Pc on the data bus. It is also written into the cache block using a least recently used (LRU) block selection criteria. Once the data value has been placed on the bus to Pc, the cache signals completion, waits for the processing unit to remove its request, and then removes the completion signal.

To write a memory location, the processing unit puts the address and data values onto the appropriate buses and requests a write operation. For each block with valid address and data fields, the requested address is compared with the block address. If a match is found, the cache entry is invalidated. The data value is always written into the memory unit. Finally, the cache signals completion, Pc removes its request, and the cache removes its completion signal.

Physically, the cache design may not be more than 6500 microns (approximately 250 mils) on a side. This size will insure good production yields. No more than 40 pins may be used for external communication.

The Structural Hierarchy

Cache memories are often described and implemented using associative or content addressable memories (CAM.) A logic-enhanced memory such as a CAM is suitable for the VLSI implementation of the cache because it is a dense and highly regular structure with a small number of special circuit cells repeated many times. This approach was adopted for the cache design even though a CAM primitive datatype did not yet exist in the experimental technology database and it was known that a significant amount of design time would be spent producing and testing a CAM template.

The top level of the cache system is shown in figures 26 and 27. It consists of the four defined datatypes "A-if," Pc-if," "M-if," and "Cache." The three

Figure 26. Encapsulation "Cache-memory"

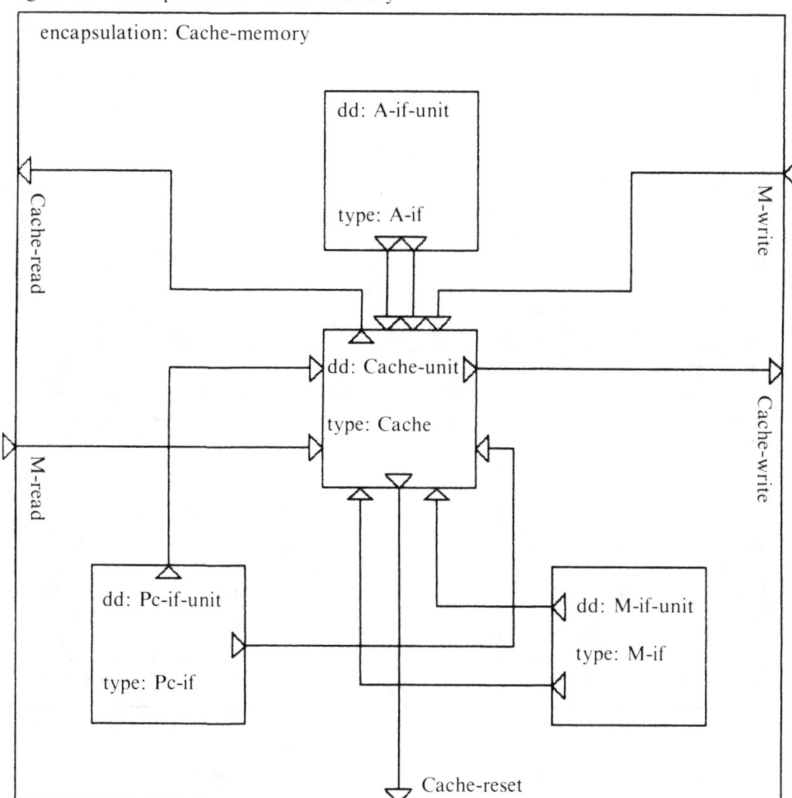

Figure 27. Floor Plan for "Cache-memory"

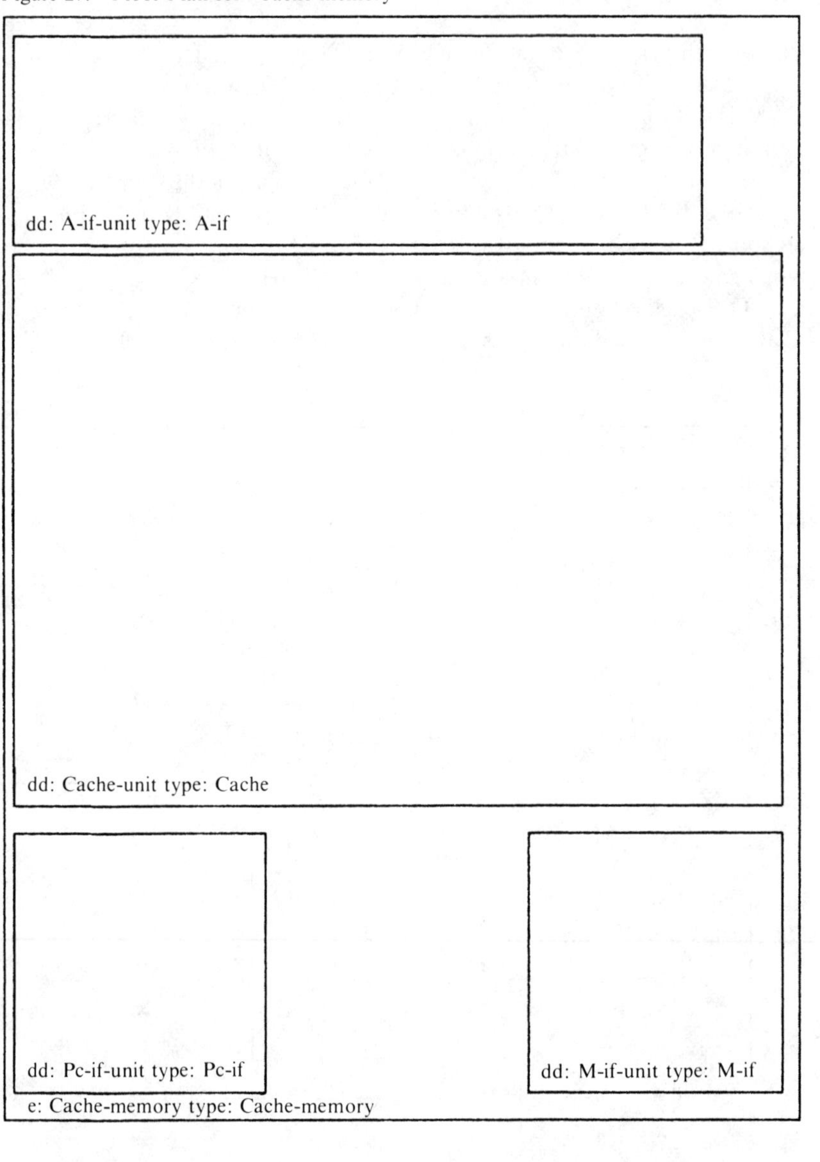

defined datatypes with the suffix "-if" implement the three major external interfaces:

1. A-if: The address bus interface. Input only.

2. Pc-if: The processing unit interface. Input and output (tristate).
3. M-if: The memory unit interface. Input and output (tristate).

"Cache" performs the data caching operations through the interface functions "Cache-read," "Cache-write," and "Cache-reset." The functions "M-read" and "M-write," that evoke memory read and write operations in the external memory unit, are called by "Cache-memory." Data values are passed between the cache and the memory unit through the defined datatype "M-if," between the cache and the processor through Pc-if," and between the address bus and the cache through "A-if."

The address bus interface "A-if" consists of an instance of the primitive datatype "ExtInput," the arbiter "Distribute," and the two defined functions "Get-Addr-A" and "Get-Addr-B" (see figs. 28-30). "ExtInput" contains the bonding pads, static protection circuitry, and input registers to receive and hold 16 address bits. Through calls to "Get-Addr-A" and "Get-Addr-B," the cache unit reads the address bus. "Distribute" ties the data and control signals together without arbitration. It is assumed that "Get-Addr-A" and "Get-Addr-B" will not make simultaneous requests. "M-if" and "Pc-if" have a similar structure (see app. F).

Figure 31 shows the structural diagram and floor plan for the defined datatype "Cache." It consists of the four defined functions "Cache-read," "Memory-read," "Cache-write," and "Cache-reset," and an instance of the primitive datatype "CAM."

The content addressable memory, "CAM," is divided from left to right into three functional units: the pattern matcher, the validity bits, and the data bits (fig. 32). Figure 33 contains the schematic for an array of four pattern bits from the pattern matcher. Each bit has a flip/flop (F/F) for storage and some gates to compare the state of the F/F with the incoming pattern. The pattern is distributed to the array by the vertical bus wires labeled "Pattern." The first bit compares its own state with the incoming pattern and passes the result of the comparison to its neighbor on the right. The neighbor forms the logical AND of this result and the truth value of its own comparison and sends this result to its neighbor. In this way, the matching process ripples through the array from left to right. With a gate delay time of 10 nanoseconds, the "Match" outputs are valid 60 nanoseconds after a pattern is driven into the array. The pattern matcher in the cache memory CAM has 4 rows of 16 bits.

The schematic for a 2-bit portion of the CAM validity bits is shown in figure 34. Each row of the CAM has one bit indicating the validity of the information stored in the pattern and data portions of that row. When performing a match operation, the result of the pattern match is passed to the validity section where it is AND'd with the validity bit. If the bit is set and a

Figure 28. The Address Bus Interface "A-if"

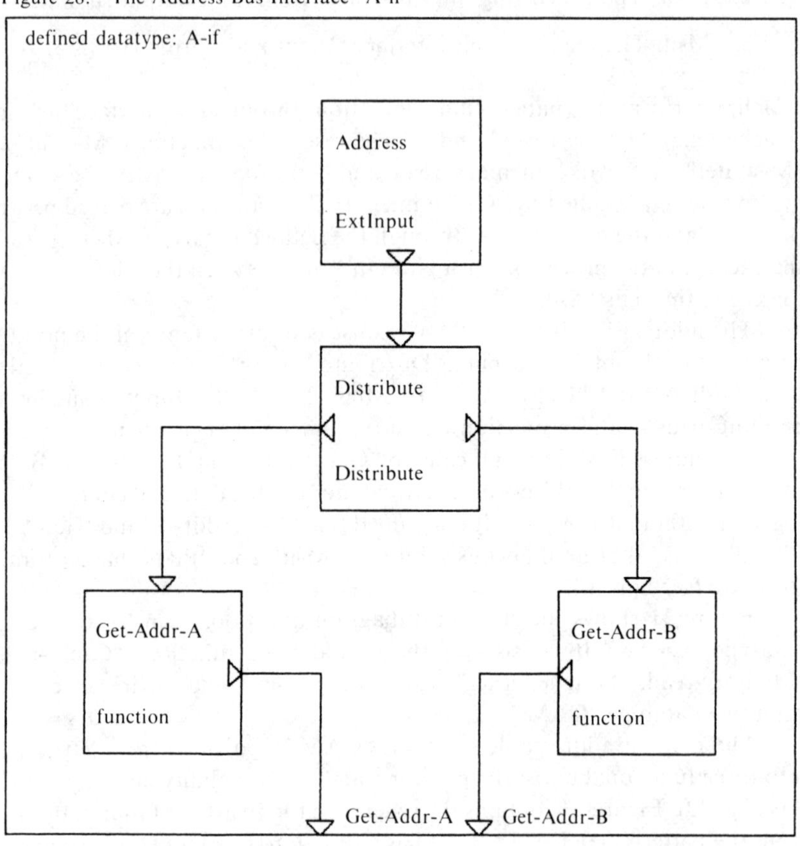

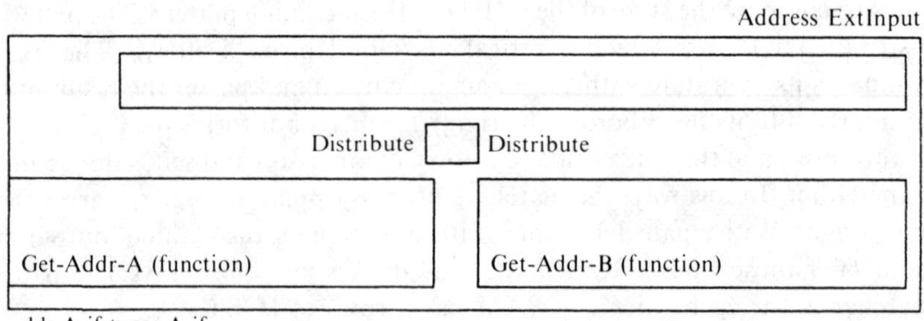

Figure 29. Defined Function "Get-Addr-A"

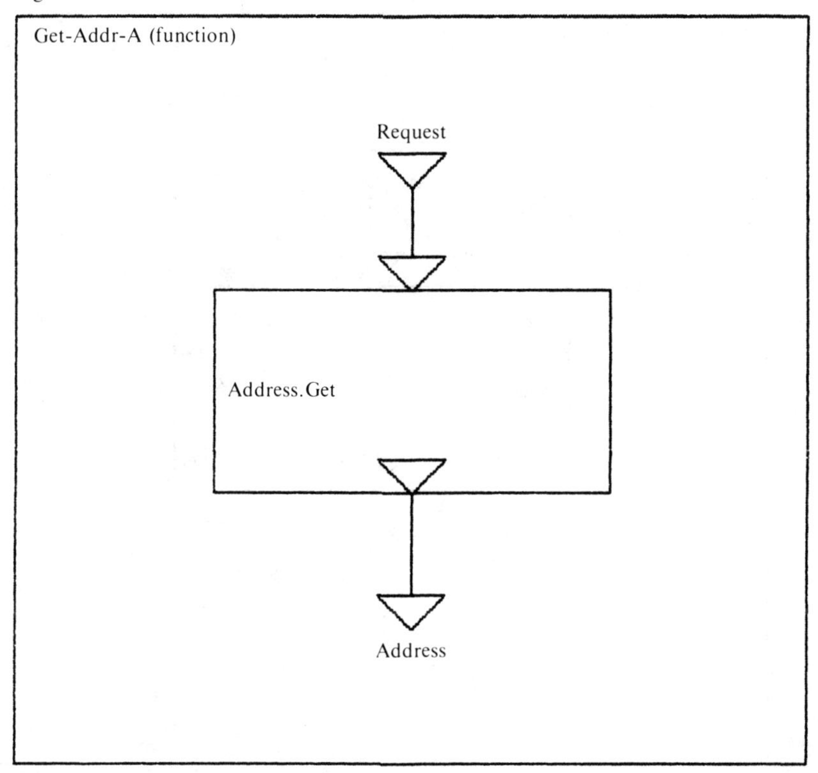

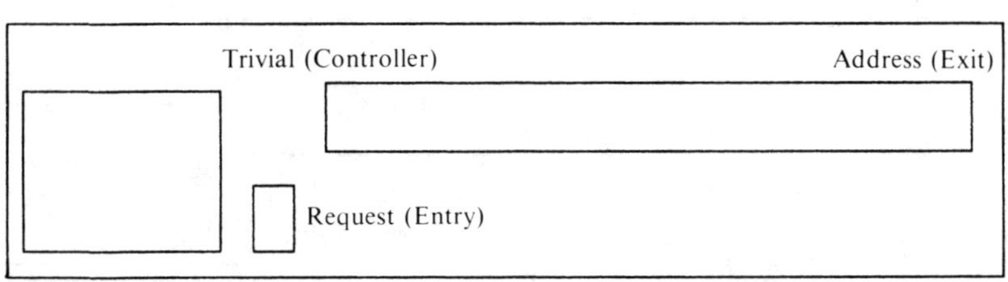

Figure 30. Defined Function "Get-Addr-B"

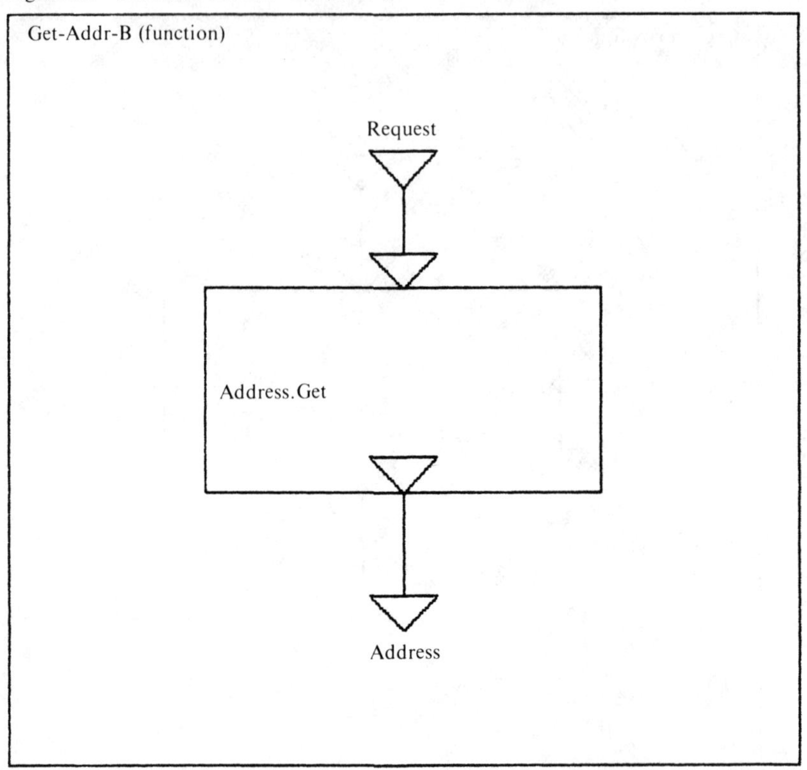

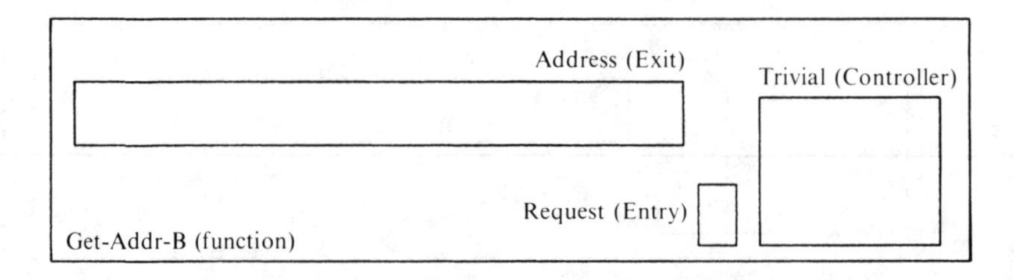

Figure 31. Structural Diagram and Plan for "Cache"

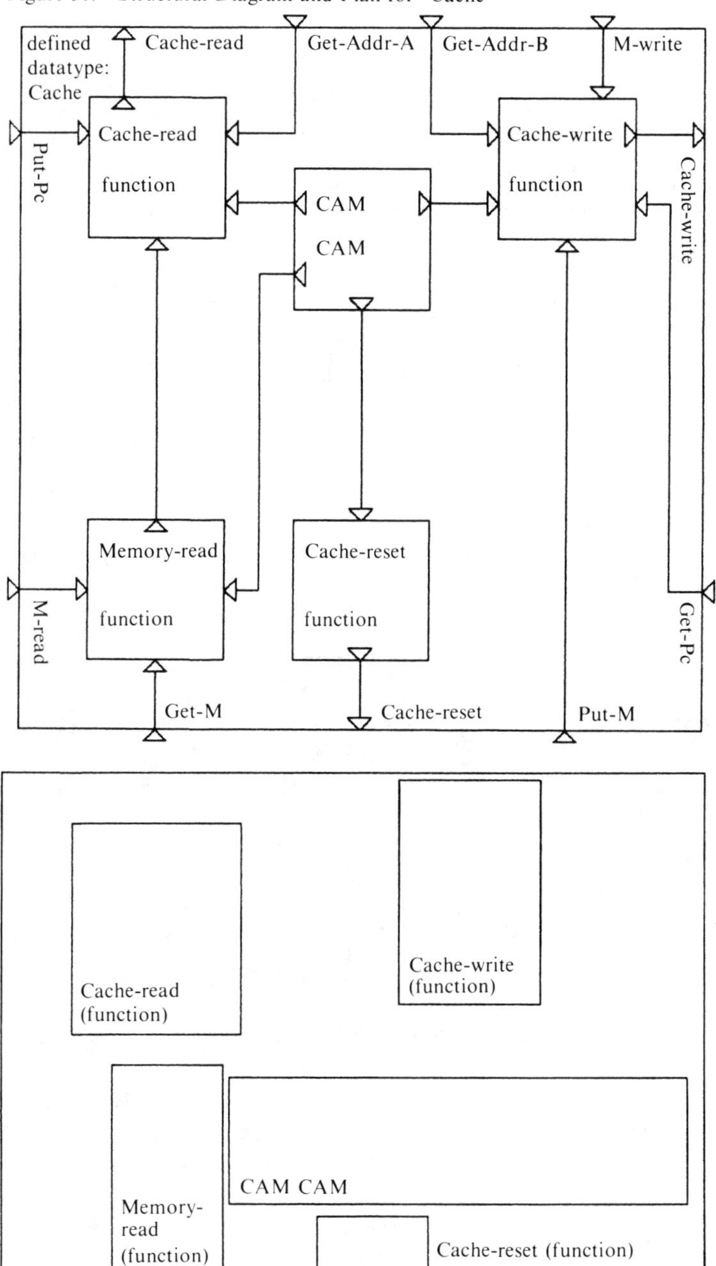

dd: Cache type: Cache

Figure 32. CAM Organization

0	Pattern_0	V_0	Data_0
1	Pattern_1	V_1	Data_1
2	Pattern_2	V_2	Data_2
3	Pattern_3	V_3	Data_3

Figure 33. A 2 × 2 Portion of the CAM Pattern Matcher

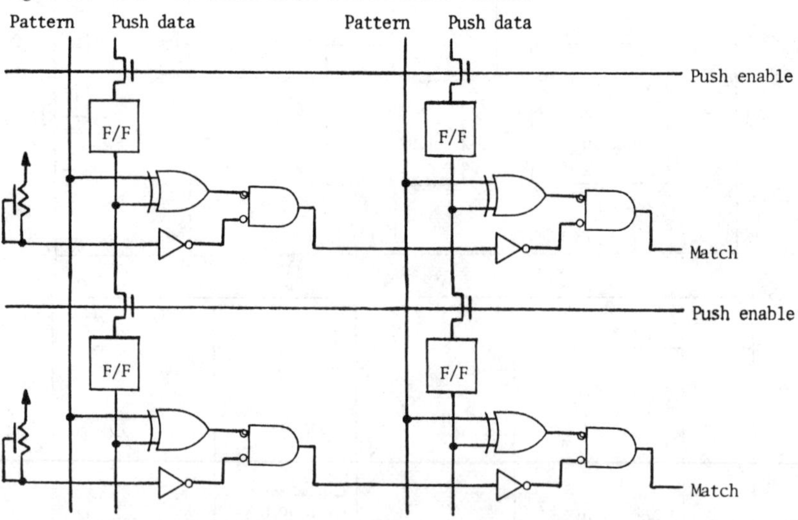

match is asserted, the "Match" output of the validity section is asserted and sent to the CAM data section. The validity bits also respond to two other signals, "Reset" and "Invalidate." When "Reset" is asserted, all the bits will be cleared. "Invalidate" selectively resets a validity bit depending upon the current state of the match signal from the pattern portion of the row. The CAM in the cache design has 4 validity bits.

During a table lookup operation, a row in the data section of the CAM will pass its contents to the CAM value outputs when the incoming "Match" bit is asserted (see fig. 35). The data bits section of the cache CAM has 4 rows of 8 bits.

Figure 34. A Two-Bit Portion of the CAM Validity Bits

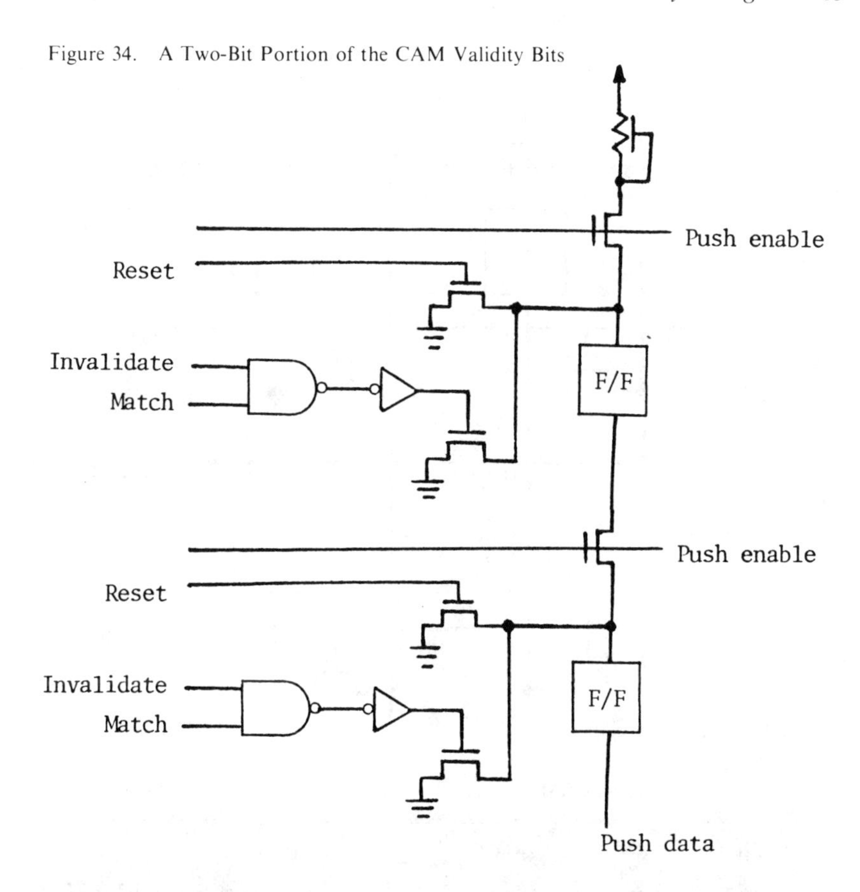

To incorporate the LRU replacement scheme, the CAM acts as a push-down array and can shift one complete row (pattern, validity bits, and data field) into another. All data for a push operation enters the array from the top. A pass transistor is placed on the input of each F/F to control data loading. The transistor gates are connected to a common bus which enables the transfer of information into the array and from one row to another row.

A truncated version of the CAM layout is portrayed in figure 36. It is a two-row, content addressable memory with two pattern bits and two data bits per row.

The Leaves

This section discusses the four defined functions "Cache-read," "Memory-read," "Cache-write," and "Cache-reset." "Cache-reset" is the simplest (fig. 37). It calls the primitive datatype function "CAM.Reset" to clear the CAM validity

Figure 35. A 2 × 2-Bit Portion of the CAM Data Bits

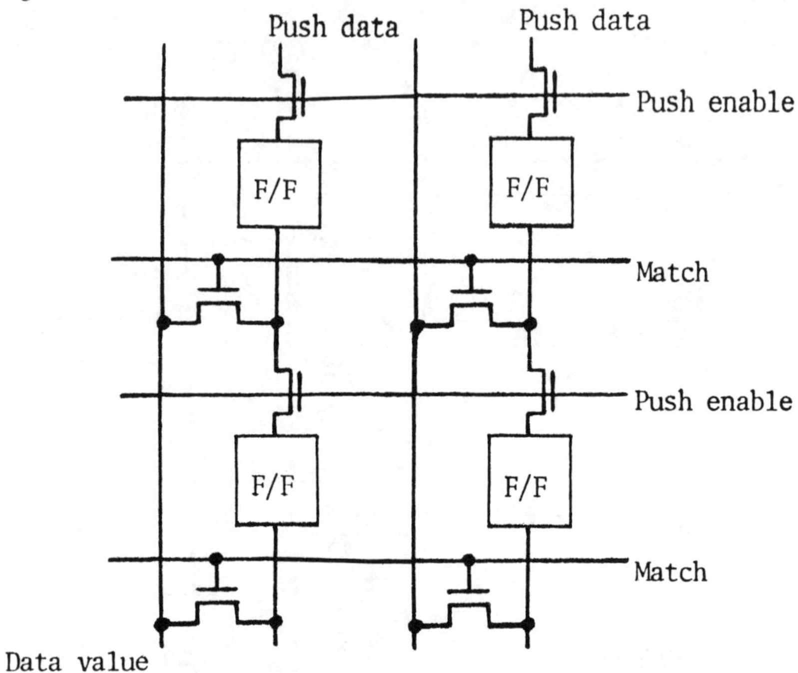

Figure 36. Truncated CAM Layout

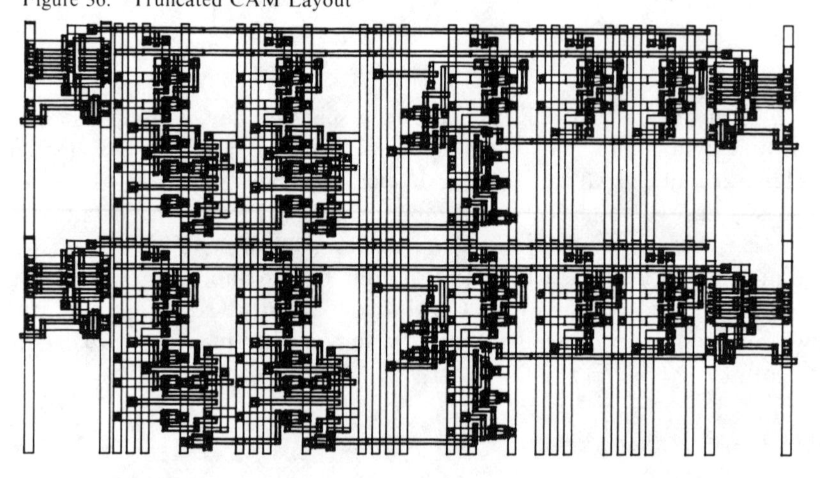

bits. "Cache-reset" is called by the processing unit as part of a system initialization or power-up sequence.

Figure 37. Diagram and Plan for "Cache-reset"

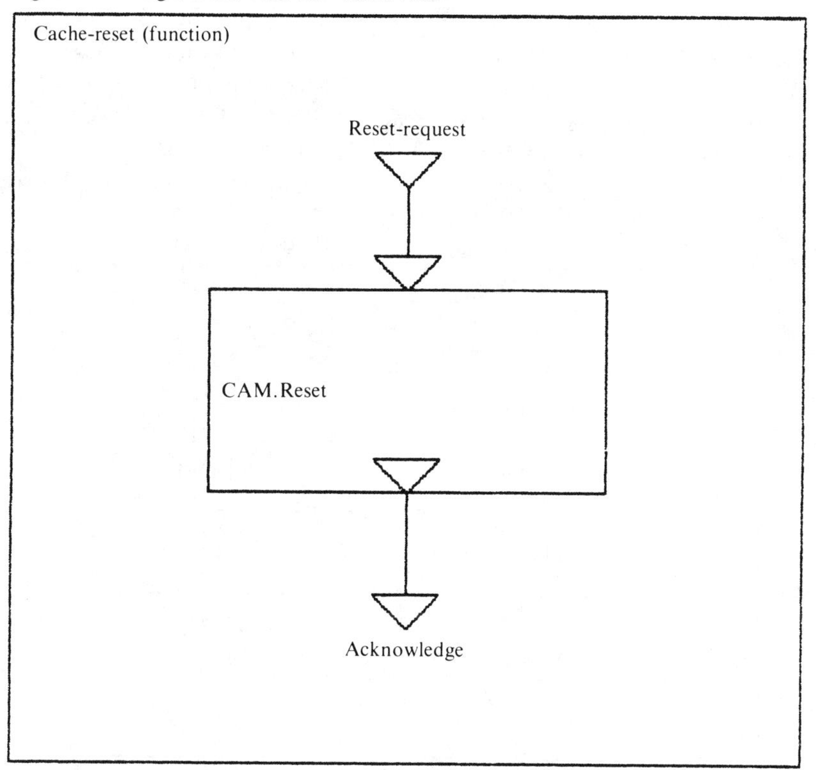

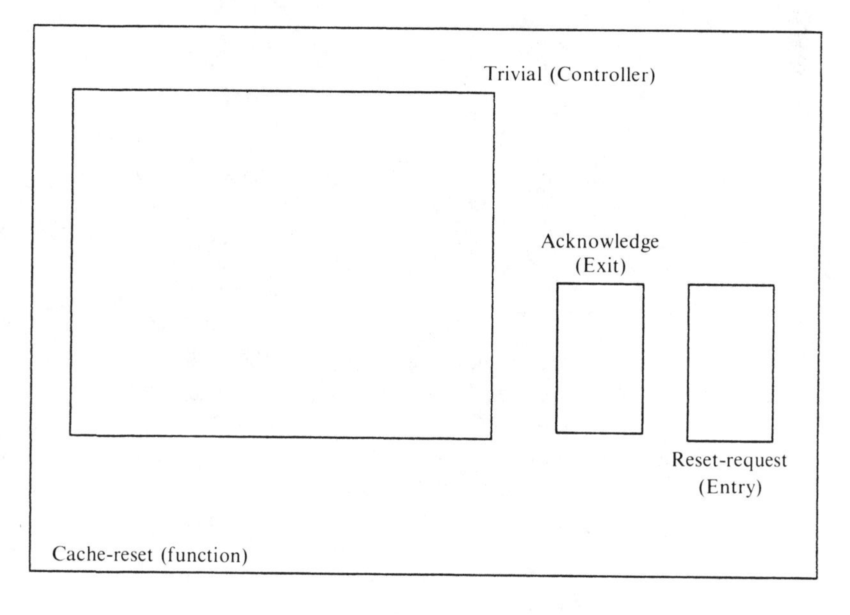

"Cache-write" is called by the processing unit when it wants to write a data value to memory (figs. 38, 39). The function has two concurrent execution paths. The left path reads the memory address from the address bus via "Get-Addr-B" and then invalidates any CAM row whose pattern field matches the memory address. The right path reads the data value from the processor interface (via "Get-Pc"), puts the value to the memory unit interface (via "Put-M"), and then invokes a write operation in the memory unit (via "M-write").

Figure 38. Defined Function "Cache-write"

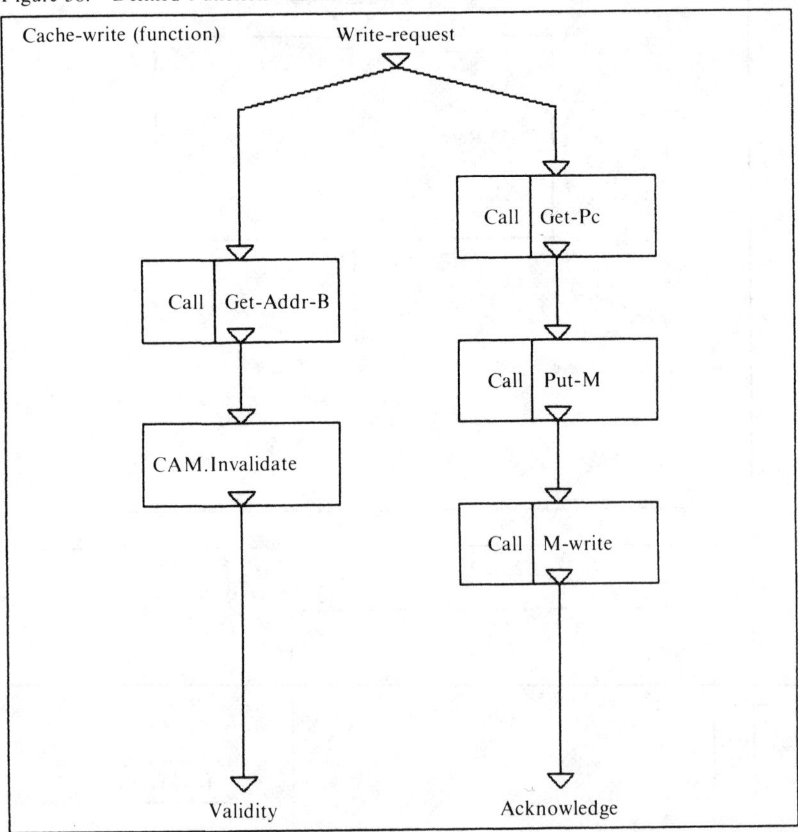

"Cache-read" reads the memory address from the address bus through a call to "Get-Addr-A." (See figs. 40, 41.) By calling "CAM.Lookup," it checks the cache for an entry with that address. The call to "CAM.Lookup" returns the data value (if found) and a bit signalling the presence of a hit. This bit is routed to two destinations. To the right, it is complemented by "Not" and sent to the "Conditional" operator which will call the defined function "Memory-

Figure 39. Floor Plan for "Cache-write"

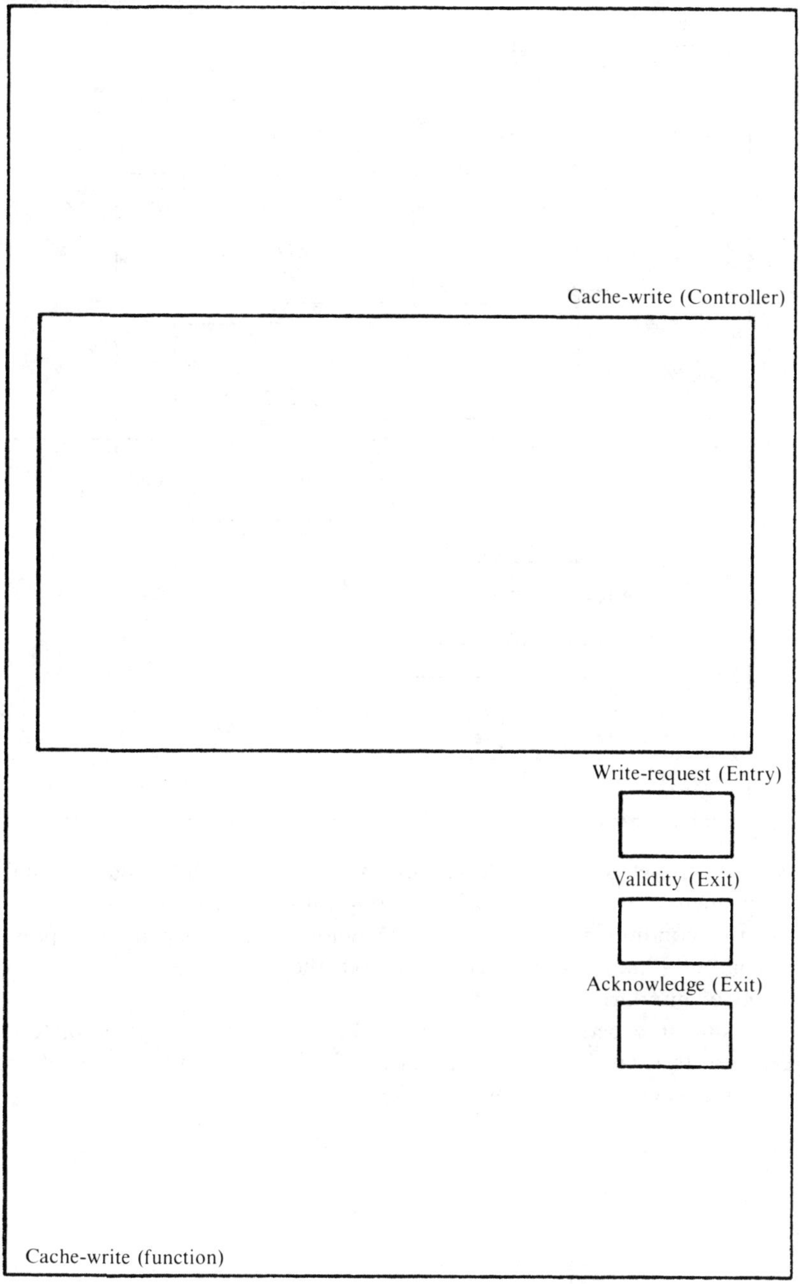

Cache-write (Controller)

Write-request (Entry)

Validity (Exit)

Acknowledge (Exit)

Cache-write (function)

Figure 40. Defined Function "Cache-read"

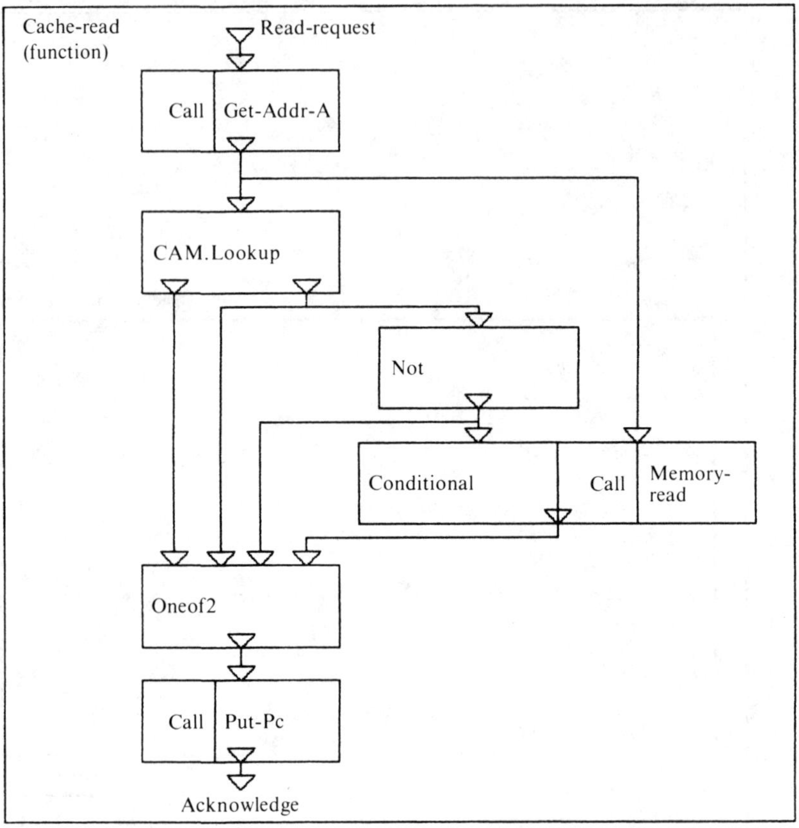

read" whenever a miss is detected. To the left, the hit condition and its complement are routed to "Oneof2" along with the result of "CAM.Lookup" and the conditional execution of "Memory-read." Depending upon the presence of a hit or a miss, "Oneof2" selects the data value to be written to the processor interface bus via "Put-Pc."

Using an experimental version of d-FSM, the function graph of "Cache-read" was translated to its state table (fig. 42). The state numbers are related to the graph nodes in the following way:

State	**Node**
1	Call Node "Get-Addr-A"
2	Primitive Datatype Function Node "CAM.Lookup"
3	"Conditional" Operator Node
4	Call Node "Put-Pc"

Figure 41. Floor Plan for "Cache-read"

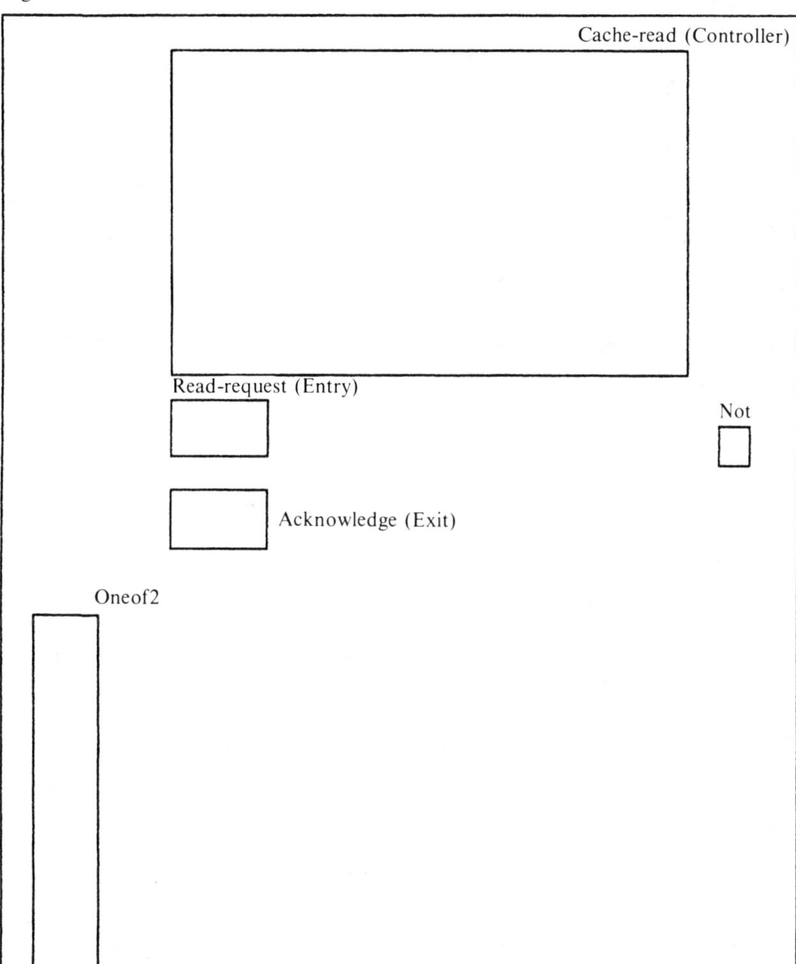

The state table was translated by hand to a symbolic Programmable Path Logic (PPL) description of the controller. This description was translated mechanically by a PPL compaction tool, arriving at the PPL layout for the controller [PetPC]. (The intermediate form and PPL for "Cache-read" can be found in app. F.)

"Memory-read" invokes a read cycle in the memory unit by calling "M-read" (see figs. 43, 44). It reads the incoming data value from the memory bus with "Get-M" and returns the value to its caller, "Cache-read." The state table for "Memory-read" is shown in figure 45.

Figure 42. State Table for "Cache-read"

```
States: (Initial Entry Exit 1 2 3 4)

Transitions:

    Initial -> Entry : (Call)
    Exit -> Initial : (not Call)
    4 -> Exit : Ack
    2 -> 4 : (Synch (2 3))
    3 -> 4 : (Synch (2 3) Ack)
    1 -> 2 : Ack
    1 -> 3 : (Synch (1 2) Ack)
    2 -> 3 : (Synch (1 2))
    Entry -> 1 :  -
```

Figure 43. Defined Function "Memory-read"

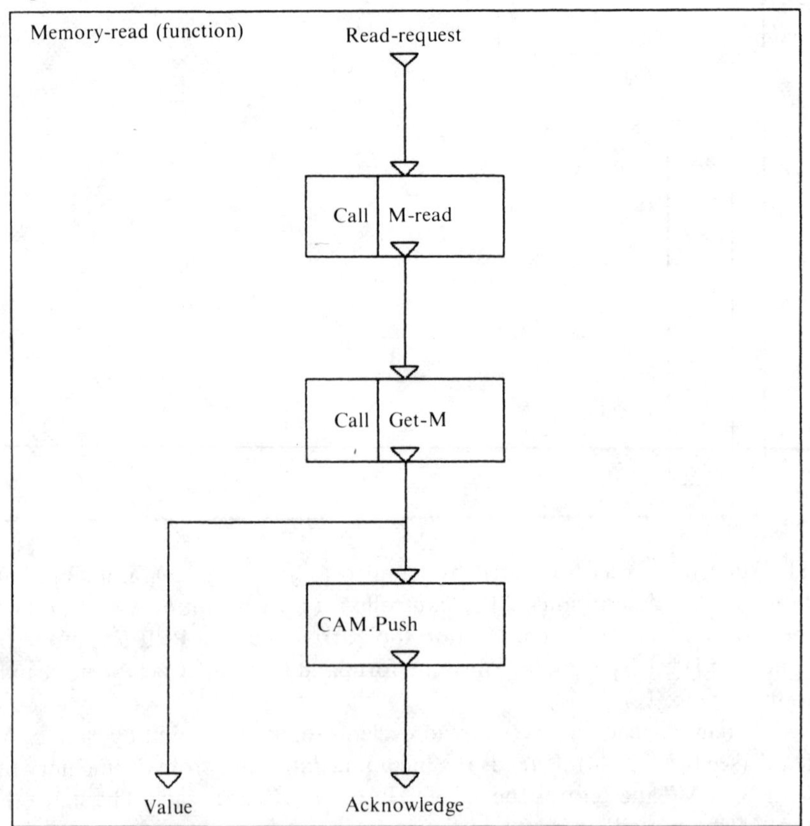

Figure 44. Floor Plan for "Memory-read"

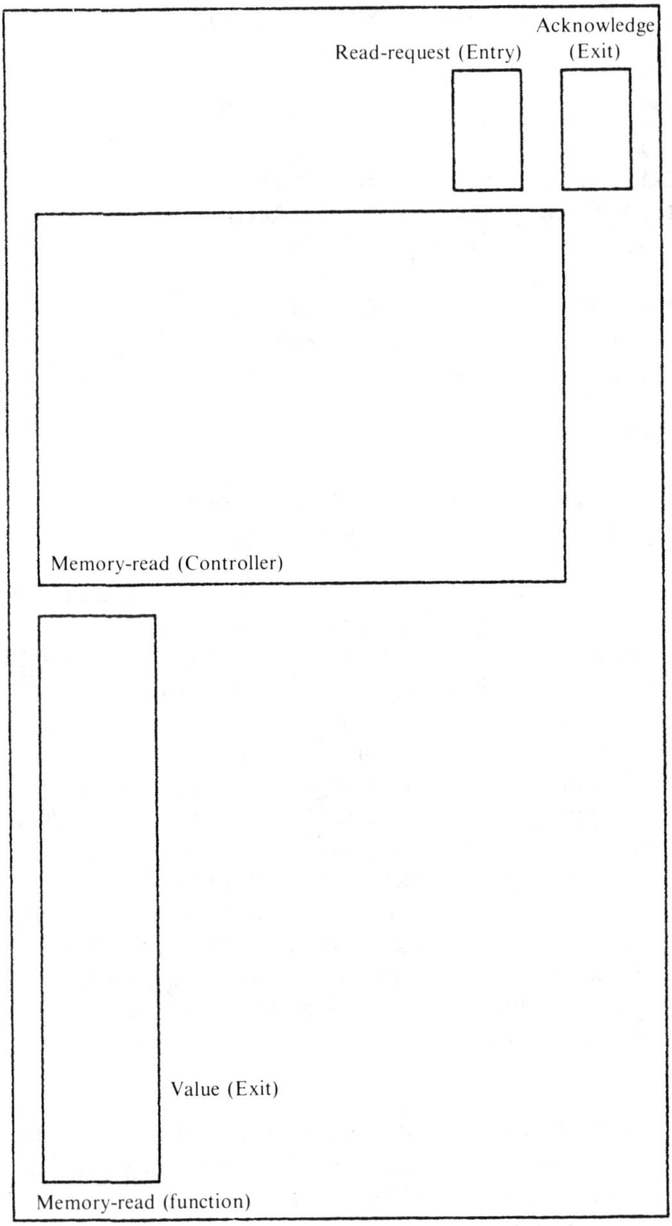

Acknowledge (Exit)

Read-request (Entry)

Memory-read (Controller)

Value (Exit)

Memory-read (function)

Figure 45. State Table for "Memory-read"

```
States: (Initial Entry Exit 1 2 3)

Transitions:

    Initial -> Entry : (Call)
    Exit -> Initial : (not Call)
    2 -> Exit : (Synch (2 3) Ack)
    3 -> Exit : (Synch (2 3))
    1 -> 2 : Ack
    2 -> 3 : (Synch (2 Entry) Ack)
    Entry -> 3 : (Synch (2 Entry))
    Entry -> 1 :  -
```

Implementation

The expanded floor plan for "Cache-memory" is shown in figure 46. The cache chip is 4800 microns wide and 6400 microns long which is within the maximum size requirement. The function inputs and outputs, primitive functions, primitive datatypes, external signals, and controllers have been mapped to their floor plan representation—a bounding box with connection points. The defined datatype and function boundaries have been drawn to establish a context for the various circuit elements. At this stage, the design is ready for delay analysis, simulation, and wire routing.

Figures 47 and 48 show the external import-export, function receptor (instance), and function call records which are the connection database for "Cache-memory." These records are created by d-router to match a function call with its receptor—a particular instance of some defined function. The matches are made through the connection identifier with each import, export, call, or receptor. The placement field in the call and receptor records specifies the absolute position of the function caller or instance with respect to the chip origin. This information is used by the delay analyzer to estimate the wire capacitances between a function call and its destination.

Delay Analysis

An experimental delay analyzer, d-da, was written in LISP and applied to "Cache-read," "Cache-write," and "Cache-reset." Sample dialogues between d-da and the designer are presented in figures 49 through 52.

The analysis for "Cache-read" was performed twice. In the first dialogue, the function was analyzed without taking wire capacitance into account (fig. 49). During the second analysis of "Cache-read," wire capacitance was

Figure 46. Expanded Floor Plan for "Cache-memory"

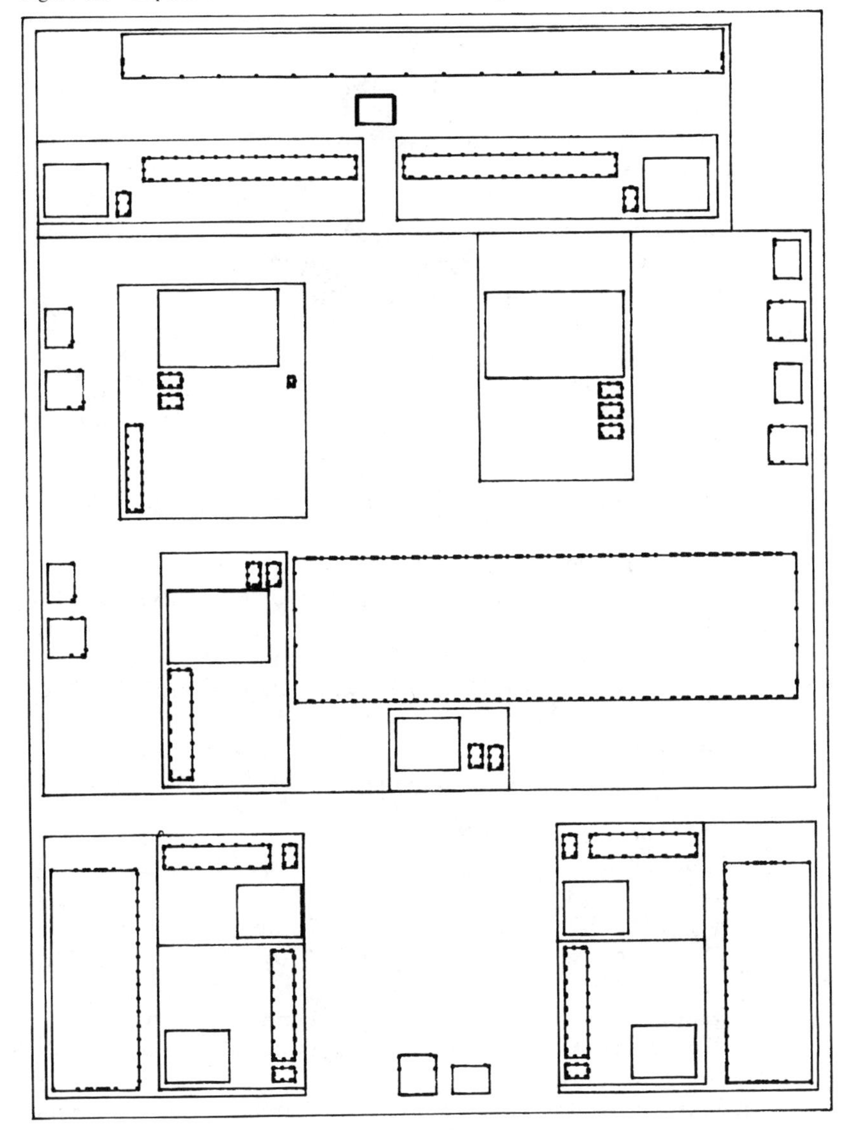

incorporated into the delay calculations. Wire lengths were estimated in the following way:

—The wire length between two nodes within a function is the Manhattan distance between the lower left corner of the first node and the second.

Figure 47. Imports, Exports, and Function Receptors

```
External imports:

    Function: M-read ID: G0167
    Function: M-write ID: G0168

External exports:

    Function: Cache-read ID: G0169
    Function: Cache-write ID: G0170
    Function: Cache-reset ID: G0171

ID: G0172
    Receptor DD: A-if              Function: Get-Addr-A
    Placement: [1.0,1.0,1.0,0.0,0.0,1.0,80.0,5200.0]
ID: G0173
    Receptor DD: A-if              Function: Get-Addr-B
    Placement: [1.0,1.0,1.0,0.0,0.0,1.0,2240.0,5200.0]
ID: G0176
    Receptor DD: M-if              Function: Get-M
    Placement: [1.0,1.0,1.0,0.0,0.0,1.0,3160.0,160.0]
ID: G0175
    Receptor DD: M-if              Function: Put-M
    Placement: [1.0,1.0,1.0,0.0,0.0,1.0,3160.0,1000.0]
ID: G0174
    Receptor DD: Pc-if             Function: Get-Pc
    Placement: [1.0,1.0,1.0,0.0,0.0,1.0,760.0,160.0]
ID: G0177
    Receptor DD: Pc-if             Function: Put-Pc
    Placement: [1.0,1.0,1.0,0.0,0.0,1.0,760.0,1000.0]
ID: G0169
    Receptor DD: Cache             Function: Cache-read
    Placement: [1.0,1.0,1.0,0.0,0.0,1.0,560.0,3480.0]
ID: G0170
    Receptor DD: Cache             Function: Cache-write
    Placement: [1.0,1.0,1.0,0.0,0.0,1.0,2720.0,3680.0]
ID: G0171
    Receptor DD: Cache             Function: Cache-reset
    Placement: [1.0,1.0,1.0,0.0,0.0,1.0,2160.0,1880.0]
ID: G0189
    Receptor DD: Cache             Function: Memory-read
    Placement: [1.0,1.0,1.0,0.0,0.0,1.0,800.0,1920.0]
```

Figure 48. Function Calls

```
Call ID: G0172
    Calling DD: Cache Function: Cache-read ID: G0169
    Function called: Get-Addr-A
    Placement: [1.0,1.0,1.0,0.0,0.0,1.0,560.0,3480.0]
Call ID: G0189
    Calling DD: Cache Function: Cache-read ID: G0169
    Function called: Memory-read
    Placement: [1.0,1.0,1.0,0.0,0.0,1.0,560.0,3480.0]
Call ID: G0177
    Calling DD: Cache Function: Cache-read ID: G0169
    Function called: Put-Pc
    Placement: [1.0,1.0,1.0,0.0,0.0,1.0,560.0,3480.0]
Call ID: G0173
    Calling DD: Cache Function: Cache-write ID: G0170
    Function called: Get-Addr-B
    Placement: [1.0,1.0,1.0,0.0,0.0,1.0,2720.0,3680.0]
Call ID: G0174
    Calling DD: Cache Function: Cache-write ID: G0170
    Function called: Get-Pc
    Placement: [1.0,1.0,1.0,0.0,0.0,1.0,2720.0,3680.0]
Call ID: G0175
    Calling DD: Cache Function: Cache-write ID: G0170
    Function called: Put-M
    Placement: [1.0,1.0,1.0,0.0,0.0,1.0,2720.0,3680.0]
Call ID: G0168
    Calling DD: Cache Function: Cache-write ID: G0170
    Function called: M-write
    Placement: [1.0,1.0,1.0,0.0,0.0,1.0,2720.0,3680.0]
Call ID: G0167
    Calling DD: Cache Function: Memory-read ID: G0189
    Function called: M-read
    Placement: [1.0,1.0,1.0,0.0,0.0,1.0,800.0,1920.0]
Call ID: G0176
    Calling DD: Cache Function: Memory-read ID: G0189
    Function called: Get-M
    Placement: [1.0,1.0,1.0,0.0,0.0,1.0,800.0,1920.0]
```

—The wire length between a calling function and the receptor of the call is the Manhattan distance from the lower left corner of the caller to the lower left corner of the receptor.

Since the wire material was unknown, a capacitance of 1.0E-16 farad per square-micron was assumed. Assuming that a wire is 4 microns wide, the estimated capacitance per linear micron is 4.0E-16 farad.

The comments below follow d-da as it threads its way through "Cache-read." (These notes are keyed to the various segments of the delay analysis dialogues in figs. 49, 50.)

1. The delay analysis of "Cache-read" starts here.
2. It first evaluates the delay of "Get-Addr-A," a self-contained function.
3. The delay of the self-contained function "Put-Pc" is then calculated.
4. The Conditional operator is encountered. The designer selects the "true" branch—the case to be executed when a miss has been detected.
5. The delay analysis of "Memory-read" starts here.
6. "M-read" is an external function whose delay must be estimated and supplied by the designer.
7. "Get-M" is analyzed.
8. The analysis of "Memory-read" is rejoined and ends here.
9. "Cache-read" is rejoined with the analysis of "Put-Pc."
10. The experimental version of d-da does not save intermediate results and starts to reevaluate some portions of the "Cache-read" graph.
11. d-da rejoins and concludes its analysis of "Cache-read" here.

The delay analysis dialogue for "Cache-write" is presented in figure 51. This function calls "Get-Addr-B," "Put-M," and the external function "M-write." Its execution time is determined by the longest (execution) time path through the controller. In this case, the longest time path passes through the call to the external function "M-write."

The delay analysis dialogue for "Cache-reset" appears in figure 52. "Cache-reset" does not call any other defined functions and is self-contained. Its delay depends solely upon the length of the longest state path through the controller and the clock period.

The most noticeable result of the delay analysis is that the cache design is slow. (Why would anyone want a cache that operates several times slower than the primary memory cycle?) Implementation of each external interface as a primitive datatype within its own defined datatype proved to be a most unfortunate organizational choice in this respect. Whenever a value is read or written to an external interface, it must pass through an extraneous defined function—a function that requires four machine states, and four more clock periods, to execute. In a more sensible approach, the external signals might be directly accessible as heavily buffered function inputs and outputs. Data values could then be read or written as a side-effect of the function call and return

Figure 49. Delay Analysis Dialogue for "Cache-read"

```
(1)     Execution delay summary for Cache-read in Cache
            This function calls other defined function(s).
(2)         Execution delay summary for Get-Addr-A in A-if
                This function is self-contained.
                Clock period: 60.0
                Longest state path: (Initial Entry 1 Exit Initial)
                Transitions in path: 4
                Execution delay: 240.0
(3)         Execution delay summary for Put-Pc in Pc-if
                This function is self-contained.
                Clock period: 228.0
                Longest state path: (Initial Entry 1 Exit Initial)
                Transitions in path: 4
                Execution delay: 912.0
(4)     Information request from Conditional NIL
        Analyze true or false branch? [t f] t
(5)         Execution delay summary for Memory-read in Cache
                This function calls other defined function(s).
(6)             Function M-read is an external import.
                    Estimated execution time: 400
(7)             Execution delay summary for Get-M in M-if
                    This function is self-contained.
                    Clock period: 60.0
                    Longest state path: (Initial Entry 1 Exit Initial)
                    Transitions in path: 4
                    Execution delay: 240.0
(8)             Clock period: 60.0
                Longest time path: (Entry 1 2 3 Exit Initial)
                Execution time: 964.0
(9)         Execution delay summary for Put-Pc in Pc-if
                This function is self-contained.
                Clock period: 228.0
                Longest state path: (Initial Entry 1 Exit Initial)
                Transitions in path: 4
                Execution delay: 912.0
(10)    Information request from Conditional NIL
        Analyze true or false branch? [t f] t
            Execution delay summary for Memory-read in Cache
                This function calls other defined function(s).
                Function M-read is an external import.
                    Estimated execution time: 400
                Execution delay summary for Get-M in M-if
                    This function is self-contained.
                    Clock period: 60.0
                    Longest state path: (Initial Entry 1 Exit Initial)
                    Transitions in path: 4
                    Execution delay: 240.0
                Clock period: 60.0
                Longest time path: (Entry 1 2 3 Exit Initial)
                Execution time: 964.0
            Execution delay summary for Put-Pc in Pc-if
                This function is self-contained.
                Clock period: 228.0
                Longest state path: (Initial Entry 1 Exit Initial)
                Transitions in path: 4
                Execution delay: 912.0
(11)        Clock period: 206.0
            Longest time path: (Entry 1 2 3 4 Exit Initial)
            Execution time: 3108.0

        Delay Analysis concluded.
```

Figure 50. Delay Analysis of "Cache-read" with Capacitances

```
(1)     Execution delay summary for Cache-read in Cache
            This function calls other defined function(s).
(2)         Execution delay summary for Get-Addr-A in A-if
                This function is self-contained.
                Clock period: 192.79999
                Longest state path: (Initial Entry 1 Exit Initial)
                Transitions in path: 4
                Execution delay: 771.19998
(3)         Execution delay summary for Put-Pc in Pc-if
                This function is self-contained.
                Clock period: 309.59999
                Longest state path: (Initial Entry 1 Exit Initial)
                Transitions in path: 4
                Execution delay: 1238.3999
(4)     Information request from Conditional NIL
        Analyze true or false branch? [t f] t
(5)         Execution delay summary for Memory-read in Cache
                This function calls other defined function(s).
(6)             Function M-read is an external import.
                    Estimated execution time: 400
(7)             Execution delay summary for Get-M in M-if
                    This function is self-contained.
                    Clock period: 154.40000
                    Longest state path: (Initial Entry 1 Exit Initial)
                    Transitions in path: 4
                    Execution delay: 617.60000
(8)             Clock period: 320.79999
                Longest time path: (Entry 1 2 3 Exit Initial)
                Execution time: 2432.7999
(9)         Execution delay summary for Put-Pc in Pc-if
                This function is self-contained.
                Clock period: 309.59999
                Longest state path: (Initial Entry 1 Exit Initial)
                Transitions in path: 4
                Execution delay: 1238.3999
(10)    Information request from Conditional NIL
        Analyze true or false branch? [t f] t
            Execution delay summary for Memory-read in Cache
                This function calls other defined function(s).
                Function M-read is an external import.
                    Estimated execution time: 400
                Execution delay summary for Get-M in M-if
                    This function is self-contained.
                    Clock period: 154.40000
                    Longest state path: (Initial Entry 1 Exit Initial)
                    Transitions in path: 4
                    Execution delay: 617.60000
                Clock period: 320.79999
                Longest time path: (Entry 1 2 3 Exit Initial)
                Execution time: 2432.7999
            Execution delay summary for Put-Pc in Pc-if
                This function is self-contained.
                Clock period: 309.59999
                Longest state path: (Initial Entry 1 Exit Initial)
                Transitions in path: 4
                Execution delay: 1238.3999
(11)        Clock period: 770.95999
            Longest time path: (Entry 1 2 3 4 Exit Initial)
            Execution time: 7790.2400

        Delay Analysis concluded.
```

Figure 51. Delay Analysis Dialogue for "Cache-write"

```
Execution delay summary for Cache-write in Cache
    This function calls other defined function(s).
    Execution delay summary for Get-Addr-B in A-if
        This function is self-contained.
        Clock period: 378.39998
        Longest state path: (Initial Entry 1 Exit Initial)
        Transitions in path: 4
        Execution delay: 1513.5999
    Execution delay summary for Get-Pc in Pc-if
        This function is self-contained.
        Clock period: 64.800001
        Longest state path: (Initial Entry 1 Exit Initial)
        Transitions in path: 4
        Execution delay: 259.20000
    Execution delay summary for Put-M in M-if
        This function is self-contained.
        Clock period: 373.59999
        Longest state path: (Initial Entry 1 Exit Initial)
        Transitions in path: 4
        Execution delay: 1494.3999
    Function M-write is an external import.
        Estimated execution time: 400
    Clock period: 243.99999
    Longest time path: (Entry 3 4 5 Exit Initial)
    Execution time: 3105.6000

Delay Analysis concluded.
```

Figure 52. Delay Analysis Dialogue for "Cache-reset"

```
Execution delay summary for Cache-reset in Cache
    This function is self-contained.
    Clock period: 77.200001
    Longest state path: (Initial Entry 1 Exit Initial)
    Transitions in path: 4
    Execution delay: 308.80000

Delay Analysis concluded.
```

mechanism. Alternatively, the extraneous function calls could be expanded "in line" by the translation tools, thereby eliminating the function call and return overhead.

Evolution

The design presented in the preceding sections is the product of an evolutionary design process. This section is a summary of that experience.

The first design, which does not appear in this chapter, attempted to use simple standard components (e.g., registers and gates) to implement the system. To gain execution speed the system was separated into a cache manager and four concurrently executing cache blocks. Read and write requests would be funneled into the manager and then distributed to the cache blocks which would operate in parallel. This organization was quickly rejected due to the complexity of the logical interconnection network. If it is a difficult task to draw the structural diagram for a system, the wiring task will be many times harder. The content addressable memory design was adopted instead.

Initially, only three primitive datatype functions were chosen for the CAM:

—Lookup: Returns the value field associated with a specific bit pattern.

—Load: Stores a pattern and value pair into the CAM.

—Invalidate: Clears the validity bit for a particular CAM entry.

The defined functions "Cache-read" and "Cache-write" were written to see if this set of primitives was complete enough to build the cache. Since the designer was more sensitive about interconnection at this stage, the procedure to choose the least recently used element (as written into "Cache-read") was judged to be too complicated. The CAM datatype functions were modified.

The functions "Lookup" and "Invalidate" were kept intact. "Load" was replaced by "Push" which loads a new entry while shifting the other CAM entries one place in the storage array. The least recently used entry is discarded by shifting it off the end. While redesigning the CAM, the designer noted a deficiency in the initial specification—nothing was stated about system initialization. So another new primitive datatype function, "Reset," was invented.

Since the new CAM appeared to be significantly more complex than the old one, detailed design was performed on the CAM. This effort produced the schematics and layout presented in the section "The Structural Hierarchy," above. Each portion of the CAM (as well as the other circuits in the experimental technology database) was simulated using SPICE to verify its electrical behavior and to measure the worst case delays through the CAM circuitry.

Next, a complete set of logical specifications were drawn along with a preliminary floor plan for each subsystem. While drawing the structural diagrams, the number of arc crossings served as a natural metric for diagram complexity and the diagrams were rearranged until the minimum number of crossings was achieved. The chip floor plan was, at first, expanded by hand (the layout expansion tool was not yet debugged) to find any obvious layout or wiring conflicts. The structural diagrams for "Cache-memory" and "Cache," in turn, guided the layout process resulting in a floor plan that is a fairly literal translation of those diagrams. Presumably, this would reduce the effort expended by d-router or the designer during interconnection.

Several additional points about the cache implementation and its style are in order:

—Each external interface was embedded within its own primitive datatype (and defined function) to the detriment of execution speed. They should be implemented as specially buffered function inputs and outputs, directly sending and receiving their values through the function call and return mechanism. In another alternative, signals could be declared to be global and could be distributed throughout the chip. The use of global signals could be considered harmful, however, due to cross-chip signal skews and losses.

—The interfaces "ExtInput" and "ExtTristate" possess their own buffer registers to hold incoming and outgoing data values. The "Entry" and "Exit" registers to hold these values are extraneous and can be eliminated. The external data values can be distributed directly to or from the interface buffer registers.

—A function input or output sometimes does not convey a data value, but just provides additional information about sequencing. The "Entry" and "Exit" circuits resulting from nonvalue-bearing inputs and outputs can be eliminated from the implementation along with their wires.

—Strictly hierarchical space planning sometimes results in a more complicated layout or routing. In "Cache-read," the result of "Memory-read" must be tortuously routed to "Oneof2"—only to be routed from "Oneof2" to the input of "Put-Pc." The selector "Oneof2" could be placed at the "Entry" register of "Put-Pc" to simplify the wiring.

—The cache memory design is not very compact. This is partly due to the use of rectangular layout shapes which do not permit particularly efficient layout. Future versions of d and d-n should support polygonal circuit and subsystem shapes. Rectangles were chosen in order to make the design and technology databases easier to manage. There is nothing in the notation or method which would preclude the use of polygons in the layout.

These advantages are artifacts of the implementation style employed in the cache example and its technology database.

Most of the design time (75 percent) was spent on the layout, coding, and testing of the d-tech-db templates. This was expected since the success of d depends heavily upon the quality of the technology database and only a primitive set of templates had been written when the cache design was started. However, as design experience accumulates in the form of new primitive functions and datatypes, a smaller portion of the design time will be spent on cell templates. Writing the d-n specifications themselves took very little time (25 percent) even though the experimental tools were relatively unsophisticated and untested. Since the experiment ended (intentionally) before the cache was wired, the amount of effort required to make all the system interconnections is unknown. With a mechanical routing aid, d should prove to be an effective aid for VLSI design.

7

System Realization

This chapter briefly discusses the experimental software tools constructed as part of this project and makes some suggestions for the future implementation of d.

Experimental Software

Several programs were constructed during this research to assess the feasibility of a real, complete CAD system based on d-n. This section briefly discusses the experimental tools and some experience in constructing and using them. The feasibility study was conducted in two phases. In the first phase, the d-n specification editors were constructed in C language under Unix on a DEC PDP-11/44 computer. During the second phase, the control generation, hierarchy expansion, and delay analysis algorithms were coded in RLISP and tested on a DECsystem-20.

In the C language package, there are tools to edit the structural and behavioral specifications (ee and de) and tools to edit floor plans (epe and dpe). Using the technology and design databases, they interact with the designer through a CRT terminal and a Grinnell frame buffer, which has a display resolution of 512 by 480 pixels (see fig. 53). The design database contains the machine readable form of the d-n system specification. The technology database contains the procedural descriptions of the primitive function cells, control operators, primitive datatypes, and primitive datatype functions.

The technology database was stored as a set of C language or RLISP subroutines. (See app. B.) These routines were loaded with the application code for the tools to form complete, executable programs. The procedural, layer-box-wire descriptions of the cell geometry proved to be quite bulky and are not always needed by every tool. For example, some tools like the floor plan editor only need bounding box and connection point information, and do not require the bulky geometry routines. This is a severe problem on a minicomputer with a very limited virtual address space like the PDP-11/44 host used in the first phase of this experiment. Future implementations should store this

Figure 53. C Language Tools

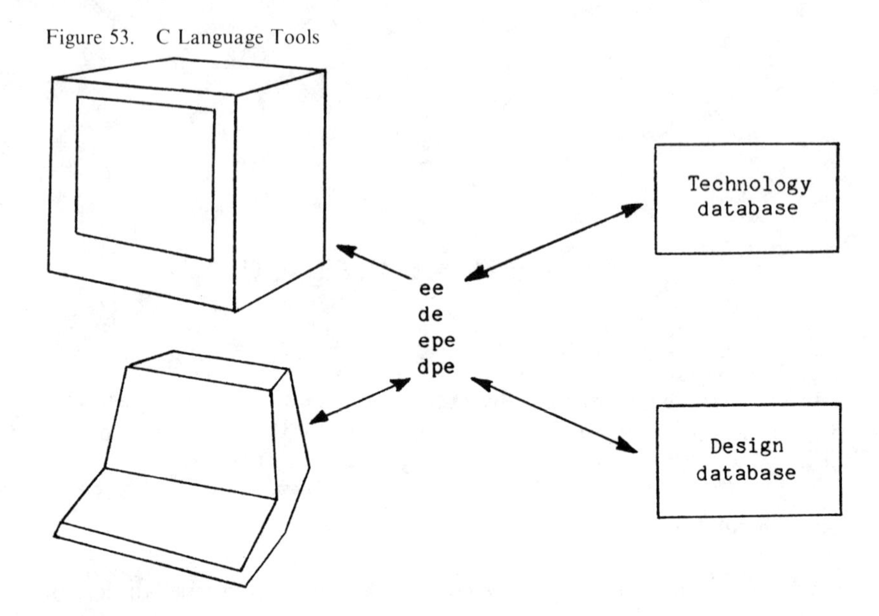

information separately so that it does not have to be loaded into every tool, whether it needs it or not.

Both C and RLISP, when suitably augmented with a graphics library, are good vehicles for cell description. C language, however, forces the programmer to grapple with too many low level programming details that are not truly relevant to list or database processing. RLISP (or some other version of LISP) appears to be a better implementation language for future versions of d because list processing and database designs are more difficult implementation problems than execution efficiency. The interpretive execution environment of RLISP coupled with its list processing capabilities made experiments much easier to conduct and shortened software development time during the second phase.

With the RLISP tools, the designer can plot d-n specifications (diagrm), generate the FSM table for a defined function (contrl), expand the hierarchy and resolve the global interconnections (hier), and calculate the expected execution time of a defined function (da). (See fig. 54.) They access the technology and design databases, although a different language and internal file format were used to construct the databases in phase two. The experimental programs are summarized in table 1 by name, size in lines of code, and function.

The RLISP tools, diagrm, contrl, hier, and da, were used to generate the cache memory example of chapter 6 and appendix F. Since these tools operate upon a design database format which is incompatible with the C language

Figure 54. RLISP Tools

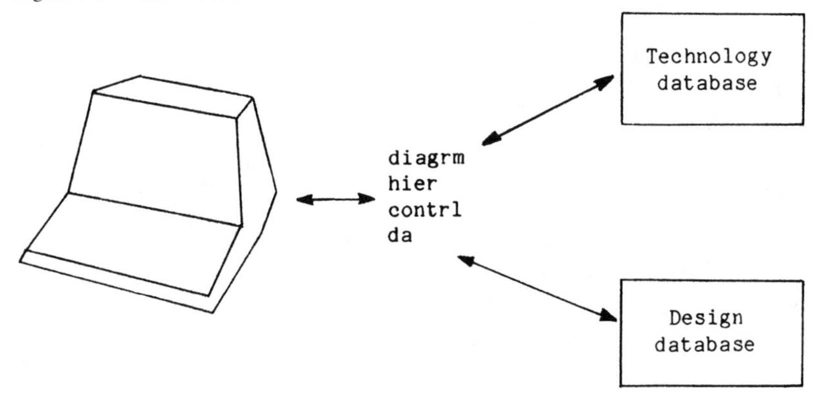

Table 1. Experimental Tools

Name	Size	Language	Function
ee	1000	C	Encapsulation structure editor.
de	1800	C	DD structure/behavior editor.
epe	1400	C	Encapsulation plan editor.
dpe	1400	C	DD floor plan editor.
db	1700	C	Design database manager.
assist	1400	C	Unix/C support routines.
tech	1900	C	Technology database.
da	250	RLISP	Delay analysis.
diagrm	850	RLISP	Diagram and plan plotting.
hier	400	RLISP	Hierarchy expansion.
contrl	400	RLISP	FSM generator.
assist	850	RLISP	RLISP/DEC-20 support routines.
tech	2000	RLISP	Technology database.

database, RLISP design databases were created using the Emacs text editor on the DEC-20. (The translation of the C language technology database to its RLISP counterpart was quite mechanical and did not pose any special problems.) The cache behavioral, structural, and floor plan diagrams were drawn with pencil and paper, digitized, and entered into the cache database. Although this was primitive, it did reduce the amount of programming needed in the second implementation phase to demonstrate the delay analysis, control generation, and expansion algorithms.

The feasibility study acted as a very useful research paradigm. In many cases, the implementation guided the development of d-n. Early experiments with different notational features and their translation were particularly beneficial. The use of textual import and export lists was one such feature that was built into the logical editors and was later rejected in favor of the purely graphical connection arcs. This proved to be a good choice since it eventually led to the development of the hierarchy expansion and global wiring algorithms. The implementation experience impressed upon the author the importance of wire management and the need for a mechanical solution to this problem.

One of the most difficult tasks during the implementation of the experimental software was the development of the design and technology databases. The amount of effort required for their development and maintenance was repeatedly underestimated. As more features were added to the editors, d-n, and other tools, the database structures were changed and sometimes the changes were radical. The incorporation of physical design information and the concept of specification versus instantiation in particular had a substantial effect on the database organization. To localize these effects, a central, custom database manager was implemented. The use of a commercial database manager would have enforced the separation of the logical and physical database structures more rigorously and would have further isolated any software modifications due to changes in the database structure. Figure 55 shows the database object hierarchy as it was at the completion of the feasibility study.

On the PDP-11/44, it was necessary to cache the design database because any given design could easily exceed the available primary memory space of a Unix process. At most, only one encapsulation, defined datatype, or defined function was accessible at any one time and, therefore, a multi-window editing style could not be supported. Further, the 512 by 480 resolution of the Grinnell frame buffer, which was used to display the graphs and floor plans, would not have been sufficient to display several levels (or views) of the design hierarchy simultaneously. This proved to be a significant drawback during interface design and layout, since the designer had to remember many assumptions about the placement of interface points or layout at the other design levels without a visual reference. A multiple window editor would require a 1024-by 1024-pixel display to support several editing windows.

The style of interaction supported by the editors was modeled after the Emacs screen editor [Sta81]. The user enters editing commands through the keyboard to position the cursor and to create or delete screen objects. Although this approach was intended to keep the designer's hands at the keyboard (instead of switching them between a tablet and the keyboard), many keystrokes are expended just moving the cursor around the screen even though a set of fast movement commands was implemented.[1] Several lessons were learned from constructing and using the editors. First, the use of the keyboard to position design elements proved to be particularly annoying when editing a floor plan due to the number of keystrokes that were needed to move an object. Instead, the tablet should be used for positioning. Next, the screen resolution

Figure 55. Database Structure for a d-n Design

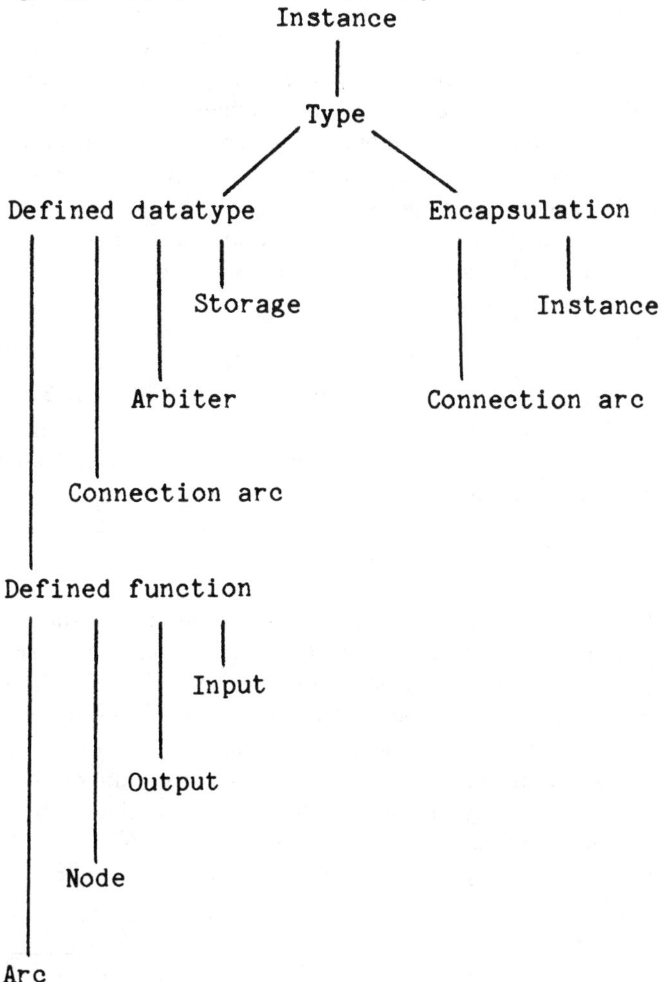

will almost always be inadequate to properly display all the detail in the physical plan of a subsystem. When connection points shrink below the size of a pixel, important detail is lost. A "zoom" feature would help this problem, allowing the user to choose an appropriate display resolution. Finally, the CAD system must provide contextual information about a subsystem (or a function) to both define and use it. The multiple window approach is suggested as a means to provide this context.

The amount of effort to construct and test a complete, correct cell description was underestimated, too. Cell geometries were first laid out on drafting paper, digitized, and coded into C or RLISP subroutines. A cell design tool modeled after REST or DPL/Daedalus should be included in any practical CAD system based on d [Mas81, Bat81]. After layout, the electrical characteristics of the cell were determined using the SPICE simulator [Spi81]. Since several simulation runs are required to formulate the worst case delay equation for a cell and it takes several more runs to debug the design, one to two days can easily be spent in the electrical design of a cell, especially if the cell geometry must be changed in response to the simulation results. A circuit extraction tool would help the designer to derive a simulation model from the cell geometry thereby speeding up the electrical design process.

Suggestions for Future Implementation

This section suggests an approach for the implementation of a CAD system based upon d. First, the general hardware and software requirements are outlined. Next, the key databases are identified and, finally, a development strategy for the software tools is presented.

Support

The proposed hardware architecture for the d CAD system is a star with a central computation and data management facility surrounded by several workstations (fig. 56). A modern time-sharing system such as the DECsystem-20 or VAX-Unix[2] coupled with a modern database management system would be a suitable host. Each of the workstations should have its own internal computer and mass storage to off-load the interactive design tasks from the central facility. A color display with at least 1024 by 1024-pixel resolution will be required to support multiple editing and tool windows. A keyboard and tablet would also be necessary for text, graph, and floor plan editing.

Figure 56. Host Architecture for d

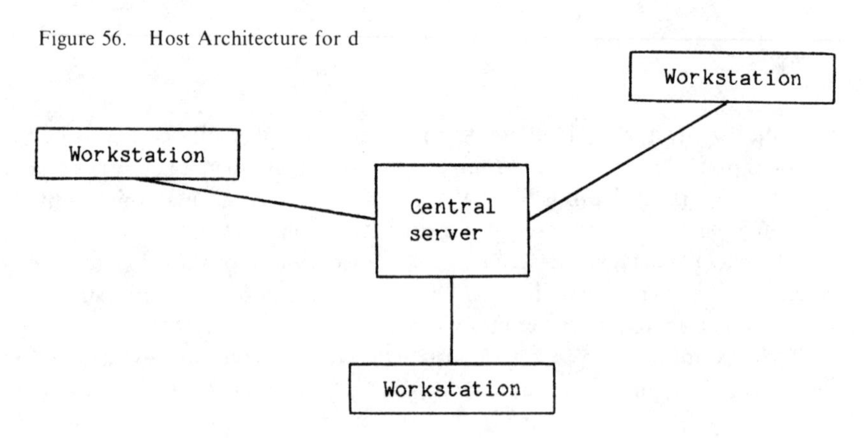

The central facility would be responsible for those services requiring a large amount of computing resources. These tasks may include, but are not limited to, database management, electrical or functional system simulation, verification, analysis, wire routing, and artwork generation. Human interface tasks, design editing, and syntactic checks should be relegated to the workstations. Through this separation of responsibilities, the central facility will be freed from frequent interruptions. Since the workstations have their own internal engine, more attention could be paid to the human interface, with the intention of making life easier for the designer. Table 2 summarizes the hardware requirements for d.

Table 2. Hardware Support Required

Workstation processor

- Low throughput, frequently interrupted computation.
- 500 kips, 256 Kbyte primary store, 5 Mbyte disk.
- Designer interaction.
- Design and cell editing.
- Delay analysis.

Workstation display

- 1024 by 1024 pixels.
- 16 colors minimum.
- Hardware assist for line drawing, text, polygon fill.

Workstation interaction support

- Keyboard for text entry.
- Tablet with puck.

Central server

- High throughput computation.
- 1-2 mips, 2 Mbyte primary store, 300+ Mbyte disk.
- Magtape drive for pattern generation tapes.
- Plotter for hardcopy of graphical displays.
- Central (project) database management.
- Functional and circuit simulation.
- Wire routing.
- Design rule checking.
- Pattern generator (PG) tape production.

Aside from a modern operating system and implementation language, the software environment must include a database manager with concurrency control. One of the primary advantages of d is the opportunity to share a design between the project engineers, both improving communication among the members of the design team and avoiding the reinvention of common subsystems and cells. Since many people will be actively reading and changing the design databases, concurrency and configuration control must be provided to prevent database and design inconsistencies.

Databases

An implementation of d will require three databases: the design database, d-design-db, the technology database, d-tech-db, and the wire or connection database, d-wire-db. Table 3 summarizes the database access requirement of the d tool set.

Table 3. Database Access

Tool	Databases Used
d-editor	d-design-db, d-tech-db
d-checker	d-design-db
d-DRC	d-design-db, d-tech-db, d-wire-db
d-cell	d-tech-db
d-FSM	d-design-db, d-tech-db
d-router	d-design-db, d-tech-db, d-wire-db
d-wire-editor	d-design-db, d-tech-db, d-wire-db
d-functional	d-design-db, d-tech-db
d-circuit	d-design-db, d-tech-db, d-wire-db
d-da	d-design-db, d-tech-db, d-wire-db
d-artwork	d-design-db, d-tech-db, d-wire-db

Tools

The construction of a CAD system requires many programmer-years of effort. To quickly bring a usable version of d on-line and to avoid unnecessary reinvention, the implementors should try to integrate exisiting VLSI design tools into the d concept of operation whenever it is possible. For example, DPL/Daedalus could be used for cell editing, electrical simulation could be performed using SPICE, and the N.mPc ISP simulator could be used for functional simulation. (Table 4 makes some other suggestions.)

Table 5 is a blueprint for the construction of d. The development team should first bring together the support software for d: the operating system, database manager, and display (window graphics) software. Next, d-editor should be implemented, giving the users a sketchpad to capture and share their designs. The translation tools, d-FSM, d-router, and d-wire-editor should then be built, followed by d-artwork. This combination of tools would permit the generation of real designs in the form of check plots and circuit masks. d-da, d-checker, and d-DRC should be constructed next to check the system timing

Table 4. Design Tools that Could be Integrated with d

Tool	Existing Tool
d-editor	New.
d-checker	New.
d-DRC	Hierarchical DRC [Whi81a, Whi81b].
d-cell	DPL/Daedalus [Bat81].
d-FSM	One-hot controller [Car81].
d-router	Channel routing [Lei81, Pin81].
d-wire-editor	EARL [DavPC].
d-functional	N.mPc (register transfer level) [Par79].
d-circuit	SPICE [Spi81], MOSSIM [Bry80b].
d-da	New.
d-artwork	CIF [Spr80].

Table 5. Construction Priorities
(highest to lowest)

Tool(s)	Capability
Support software	OS, windows, database mgt.
d-editor	Edit design database.
d-FSM, d-router, d-wire-editor	Translation.
d-artwork	Mask generation.
d-da	Delay analysis.
d-checker, d-DRC	Design checks.
d-functional, d-circuit	Simulation.
d-cell	Graphic cell editing.

and layout. This level of capability would provide some early "traps" for timing, syntactic, and layout errors. Following this step, electrical and functional simulation tools would be essential, since designers still rely upon simulation for the logical and electrical verification of their designs. Finally, the development of d-cell would ease the design of new circuit cells and would hasten the expansion of the technology database.

Of course, this list of priorities represents an ideal and will not fit every situation. For example, d-n specifications are simple enough to be drawn by hand using pencil and paper, and a graphical editor is not really necessary. The specifications could just as well be translated to their geometric implementation using an artwork editor such as CAESAR, ICARUS, or Daedalus. Many sites already have PLA or ROM generation programs. Hence, the control portion of a d design could be translated by hand into PLA or ROM programming files. A version of d-da (such as the experimental prototype) could still be employed to check timing behavior before committing the design to extensive circuit simulation since the basic data path and control delays would be known or could be calculated. The cells could be interconnected manually with the artwork editor while using the d-n structural and behavioral diagrams as a guide. Similarly, simulation models could be derived from the logical diagrams and the physical geometry.

An alternative to a full-blown, customized d system is illustrated in figure 57. In this organization, a central file manager would regulate access to the d-n specifications and cell descriptions that make up the description of the VLSI system. A collection of translators would surround the file manager and would separate the central design repository from the design tools themselves. The translators would transform the design description from its central, common form to the description format of a particular design tool. One such translator could transform a d-n specification to an electrical network in the SPICE description format, for example. In this manner, a project without a large amount of resources for the development of design tools could capitalize upon its existing investment in VLSI design tools without sacrificing some of the advantages of d and d-n—especially those of a common communication medium and design repository.

Figure 57. An Alternative to a Full d CAD System

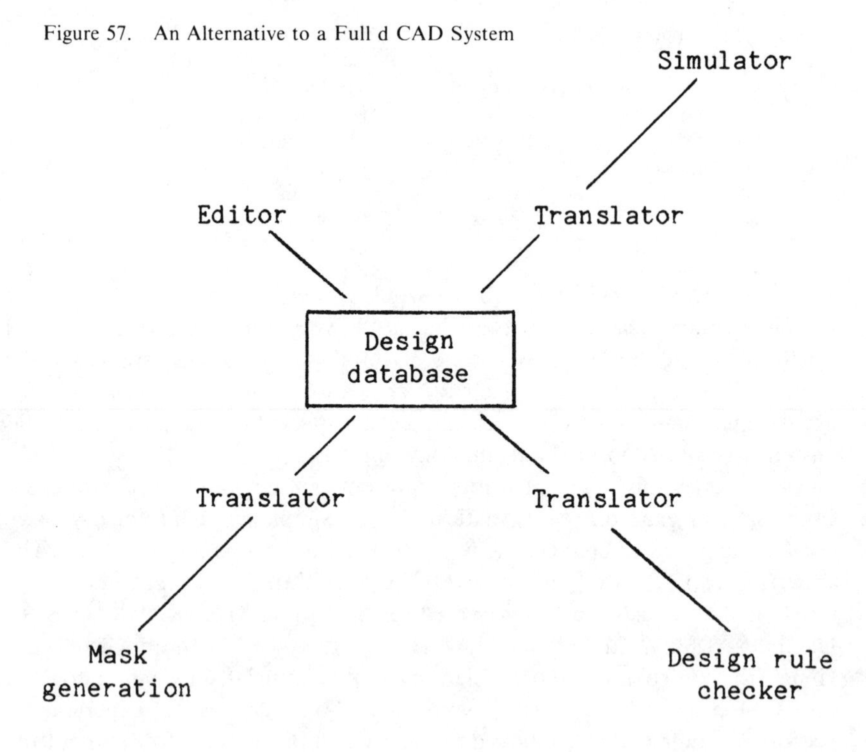

8

Evaluation

This chapter evaluates d from several perspectives. d and d-n will be judged against the primary research goals of this project as presented in chapter 2.

Linguistics

Several linguistic ideas appear throughout d-n and the descriptions written in the notation. All design objects, whether they are primitive datatypes, defined datatypes, or encapsulations, can be modeled as representational datatypes. Each level of the hierarchy is characterized by the defined functions that it defines and uses. Even the lowest level, which is defined by the procedural circuit templates in the technology database, consists of simple primitive data operations and datatypes. The same notation (graphs and floor plans) is used at every level of the hierarchy to describe the system structure, behavior, and physical arrangement. Therefore, the first primary research goal was met: d-n is a self-consistent notation for structured VLSI design.

Although the representational datatype has a certain conceptual appeal for design,[1] the use of defined functions (subroutines) to model subsystem interactions and communication results in a somewhat unorthodox view of a hardware system. Further, the linkage overhead between a caller and its receiving function was found to be a significant portion of the estimated function execution time. Typically, an interface is a place where information is quickly exchanged between two cooperating processes. d-n could (and should) be modified to adopt the more conventional process model [Bel72, Hil73].

Graphs and plans are an appropriate notation for the description of a VLSI system. Floor plans have been applied in practice for many years and their use is widely accepted. d-n helps the floor planning process because it enforces hierarchical design. The allocation and placement of subsystems and circuits track the processes of system decomposition and synthesis.

Graphs make the structure of any system or computation instantly apparent and are, therefore, particularly appropriate to the VLSI design problem. By inspection the designer can tell if a particular interconnection topology will be too complicated to lay out conveniently. When used to describe system behavior, graphs naturally represent horizontal concurrency—one of the primary benefits of a VLSI implementation. Data dependency graphs do not require any particular control architecture and can be translated to an asynchronous self-timed environment as well as a strictly sequential, clocked control organization. Descriptions which are based upon a synchronous control architecture do not readily admit self-timed, asynchronous implementations.

d-n restricts the designer in several ways. First, two defined datatypes cannot share the same memory unit directly. Shared access to a storage element must be regulated by a third defined datatype and, unfortunately, this indirect access causes an unacceptable delay in some applications. Next, cyclic subsystem definitions and recursion are prohibited. This should not pose any significant limitations unless the design uses a recursive algorithm that cannot be restated as an iterative procedure. Finally, arbitrary interconnection topologies cannot be formed because strict hierarchy must always be observed.

The description of d-n in chapter 4 (and app. A) establishes a set of translation rules for a specification from the logical design domain to the physical implementation domain. This translation is total and satisfies the second primary research goal. Every encapsulation, defined datatype, defined function, and circuit is represented by a physical structure on the chip as reflected in the floor plans. An algorithm is presented in appendix C that translates a function graph into a state table suitable for implementation as a one-hot state machine controller. From the logical interconnections and the absolute positions of the corresponding physical connection points on the chip, a complete wire list can be generated (app. D). An automatic router or a manual wire editor could be used to draw the actual wire paths between those connection points—including power and ground connections.

The lack of a well-defined target language for the logical to physical translation process is the major flaw in this work. The target is too general and it does not put enough constraints on the translation process to arrive at a clear and concise expression of the physical implementation. More structure and constraints need to be imposed on the target environment:

—To improve wire management (routing),

—To place circuit cells automatically, and

—To clarify the descriptive semantics in the logical domain.

In the case of state table generation, where the target and the translation algorithm are well defined, the understanding of defined-function semantics has been enhanced and the translation is more effective. One way to incorporate additional constraints into the translation process is to adopt a fixed target language such as Symbolic Logic Arrays (SLA) [Smi81a, Smi81b]. The SLA cell set handles local wiring, power routing, and readily supports the one-hot control scheme.

This negative result does not invalidate the notation or the method. It simply means that future work on d should carefully specify the target environment before applying the technique and its tools.

Separation of Domains

d-n separates the design of the behavioral and structural domains from the physical domain. Since virtually all of the physical layout, wiring, and controller design are suppressed in the logical specifications, the separation of domains in d-n accomplishes the third primary research goal. The logical specification style is a true linguistic approach to VLSI design. A VLSI system is no longer treated on the structural and behavioral levels as a carefully arranged set of rectangles which incidentally compute some function.

It is easy to see how the logical design drives the physical one. Subsystems, functions, and interconnections are brought into existence and must eventually be mapped to circuits and wires. But, how does the physical domain affect the logical design? There are four primary factors that affect the physical design of a VLSI system:

1. Size: The design must fit onto the chip. There is a direct relationship between size and yield. If the design is too large, the yields after fabrication will be unacceptably low.
2. Speed: If the result is late, the part is unusable.
3. Power consumption and dissipation: Excessive current densities in conductors cause metal migration, eventually destroying the circuit path. All circuits generate heat which must be dissipated to the package. If excess heat cannot be dissipated, the circuit will be damaged.
4. Testing: Since the fabrication process has many sources of error, chips must be tested to separate the good circuits from the bad. Inevitably some design errors will be made. Testing is an important part of debugging.

The problem of physical design is really a search for an acceptable balance of these factors. If a balance cannot be found, the logical design must be changed.

Wiring is a difficult task because it affects, and is affected by, all four of these factors and their interactions. Typically, the wires occupy 80 to 90 percent of the total chip area. Speed is directly dependent on wire length because each of the interconnection paths is a capacitor which must be charged according to some time constant. To keep delay constant, a longer wire requires a bigger drive transistor that increases the chip size and power consumption. To avoid metal migration problems, the current densities in the power grid must be closely monitored. Finally, wires must be used to bring out test signals, complicating the layout process still further.

In the logical descriptions, the designer makes interconnections without regard for wire length. After delay analysis or simulation has been applied to the corresponding physical design, any one of the four factors may be found wanting. If rearrangement of the physical design cannot eliminate the problem, the logical specifications must be changed. d does not abandon the designer, however, since the delay analysis and simulation tools (d-da, d-circuit, and d-functional) accelerate the feedback loop. More information about a design can be derived earlier in the design process to varying degrees of accuracy (e.g., nominal wire delays, delays based upon the Manhattan distance between two circuits, and actual wire delay). The delay estimates based upon Manhattan distances appear to be quite useful since they are easily derived by the CAD system and they do not force the designer into the time-consuming process of wire routing too early.

Design Analysis

Mechanical analysis is essential to the VLSI design process because there are so many details, factors, and trade-offs to be considered by the designer or project team. Space and time trade-offs are addressed by d, satisfying the third primary research goal.

The d-n floor plans play two different roles. First, they describe the placement of subsystems and circuits on the chip surface. The chip geometry and patterning masks can be produced from these floor plans, the wire geometry, and the procedural circuit templates stored in the technology database. Next, the floor plans provide valuable information about space utilization. From the layout, the designer can decide if a given subsystem is too large, if enough space has been allocated for wiring, or if the design is too big for implementation as a single chip.

The mechanical delay analysis incorporated in d is a new contribution to VLSI design. The execution speed of a system function can be estimated from the logical and physical descriptions. As the system is laid out and the wiring is added, better estimates for performance can be made. Since the procedural technology database model permits the description of adaptable circuit elements, the design can be reparameterized to reduce circuit delays. This will inevitably affect the size of the circuits and these changes will be reflected in the floor plans. Hence, the visibility of space and time trade-offs will be improved.

System partitioning and sizing are important steps in any VLSI development effort. The design problem must be partitioned into manageable subproblems that will fit onto one or more integrated circuits and still operate as a coordinated unit. The decomposition and sizing problem has been ignored by other tool builders, but is addressed by d. The structural diagrams are a more formal version of the block diagrams that designers now use to decompose their systems. The transformation from structural diagrams to d-n floor plans is direct and can be made very quickly using the d-editor, directly assisting the system decomposition task. Since the delay analysis algorithm can form reasonably good estimates for execution speed from the d-n descriptions (without resorting to detailed layout and wiring), the estimates can be used to make better decisions about system structure much earlier in the design process.

Pragmatics

Modern systems are beyond the complete comprehension of any single individual. For this reason, a team of engineers is usually assigned to a design project. Team efforts have their own special set of problems. If project members do not communicate with each other and their designs fall out of step, a number of incompatibilities will result:

—Interfaces are changed and become incompatible.

—A subsystem grows beyond its budgeted space requiring a change to the floor plan.

—A critical, low level function executes too slowly affecting the performance of the rest of the system.

—Two engineers unknowingly begin to design the same circuit cell or function duplicating their efforts.

—The rationale for a particular decision has been lost and the team cannot reconstruct their argument.

By sharing information about the design in a controlled way, these problems can be alleviated.

d encourages sharing by providing an unambiguous communication medium (d-n) between the engineers. The design and technology databases are a common repository for design experience. Through administrative controls embedded in the tools, unauthorized modifications to the design configuration can be prevented. Proposed changes can be evaluated against the most recent, stable version of the system. System level decisions concerning execution speed, space allocation, interfaces, and global wiring can be coordinated. By maintaining a historical record of modifications, design decisions are captured in the databases and analytical results can be reproduced.

The experimental implementation of the d-n editors, technology database, and delay analyzer indicates that a d-based CAD system can be constructed. The experimental tools do not support the multiwindow human interface discussed in chapter 5, but no insurmountable implementation problems can be foreseen. Those tools that have been developed by other groups with capabilities similar to the ones proposed for d (e.g., simulators, design rule checker, state table generator, pattern generator, and cell editor) were not constructed, but could be integrated into a real d CAD system. Hence, the fifth research goal, feasibility of implementation, has been met.

The most problematic and practical issue in VLSI design is wiring. d stakes its success on the development of a good automatic routing tool and, eventually, a tool for mechanical placement. Recent advances in routing and placement techniques give every assurance that good software tools will be available in the next two to three years [Lap80, Lei81, Pin81]. In the near term, several techniques can be exploited to generate more of the wire geometry automatically:

—Employ a standard cell package that handles local wiring and power distribution.

—Use multiple layers of metal, instead of a single layer, for interconnections. This requires a nontrivial change in the fabrication process.

—Permit nonrectangular subsystem and circuit shapes giving the designer more flexibility and higher density in the layout.

—Use circuit cells with deformable aspect ratios (pitch.) The procedural technology database can easily accommodate deformable cells.

With the advent of automatic routing, the prospects of a "silicon compiler" based on d and d-n are bright.

9

Summary and Future Directions

A method for VLSI design, its descriptive notation, and its tools have been presented. The d method provides a structured, self-consistent approach to the design of nMOS VLSI systems. Designs are expressed in the d-n notation which separates the description of behavioral and structural (logical) domain design from the physical domain. This separation keeps extraneous detail about the circuit geometry and layout from interfering with the logical design, moving the VLSI design process into the realm of programming. Through a set of translation rules, the descriptions can be mapped to real circuits. Tools to support d-based design were discussed. One of these tools uses information about the logical to physical domain translation to evaluate system performance. This analysis provides feedback to the designer about circuit delays and execution speed. Using the analytical results as a guide, the designer can change the logical and physical specifications to find a better mixture of execution speed and space utilization.

The rest of this chapter discusses several different directions for future research on d, d-n, and the tools.

Application

First and foremost, d should be applied to a real VLSI-sized problem. It has been the experience of many software engineers that techniques which work for small scale projects do not "scale up" to very large scale systems. Hopefully, the first project will have a relaxed schedule to let d evolve with the design problem.

As part of the application problem, a rigorously specified translation target should be selected (or defined). Software compilers *are* successful because their target machines are quite well defined. Many cell packages such as Symbolic Logic Arrays handle local layout and wiring. The capability to invent and create new cell designs should not be surrendered, however. No package designer can be clever enough to anticipate all possible applications and design situations.

Tools

The power of a tool depends largely upon its relationship with its user. The multiwindow display approach suggested in chapter 5 can help the designer comprehend and manipulate the design in several different abstract problem domains at once. Many other graphical aids are possible such as tracing simulated execution on the function graphs and structural diagrams while using the display as an oscilloscope to observe the circuit's electrical behavior.

The circuit templates in the experimental technology database were produced in the worst possible way. They were first drawn on drafting vellum, digitized, and then coded into C or LISP. Clearly, a cell editor would have assisted this process. If the cell geometries were static, this would not pose any significant problems since an almost excessive number of cell editors are available in the marketplace. However, an editor for geometrically adaptable cells has not yet been developed.

Extensions

Aside from logical design and layout, the key problem in every integrated circuit design project is testing. Unlike a software system whose operation can be monitored virtually anywhere within the program structure, the behavior of an integrated circuit cannot be conveniently observed. Since the fabrication process is fallible, working IC's must be separated from the bad, and this must be done quickly if the production volume is large. The application of d and d-n to the generation of test or signature patterns should be explored.

Another important extension to d is to incorporate the definition of performance constraints into the d-n descriptions. In a CAD system where the designer initiates revisions to the design, the CAD system could automatically compare the estimated execution speed of a function or subsystem against the constraints and flag those functions which will not operate within the specified bounds. If the CAD system automatically explores the design space for the designer, it will need the performance constraints to find or detect a goal state that satisfies the performance criteria.

Analytical techniques similar to the delay analysis algorithm should be explored, also. If each cell in the technology database could calculate and report its power requirement, the total power consumption of the chip could easily be estimated. Since the d-n descriptions are graphs, a complexity measure could be applied to them to determine if a graph is too complex to be successfully wired by the mechanical wire router. The designer could then modify the design and reduce the interconnection complexity.

A successful VLSI development effort depends upon the quality of its prefabrication analysis. Designers use simulation and design rule checks to detect as many errors as possible before manufacturing prototype devices. d enhances the use of these conventional tools by providing a standard, unambiguous representation (d-n) for the design and introduces several new mechanical aids into the design process. These tools help the designer to make decisions faster and with better information.

The selection of representational datatypes for d-n was not accidental. Simulation and testing can demonstrate the presence of bugs, but they cannot show their absence. It is hoped that future research on the formal verification of axiomatic (formal) datatype specifications can be applied to VLSI design. A language based on representational datatypes should provide a suitable translation target for axiomatic descriptions. As formal specification and verification technology increase in sophistication, its application to VLSI design through the d system must be explored. Only through mathematical analysis will we be able to reckon with the ever increasing complexity of integrated systems.

Epilogue

This section summarizes more recent advances in silicon compilation, hardware description languages, and digital systems synthesis.

Stanford Palladio

The Stanford Palladio design environment is a knowledge-based system for the design of digital circuits [Bro83]. Palladio is an integrated design system that provides a compatible collection of CAD tools for specification, simulation, analysis, and layout. Its implementation uses several different programming paradigms: rule-based, data-oriented, object-oriented, and logical reasoning. Palladio is implemented in the Lisp Object Orientated Programming System (LOOPS) and the Metalevel Representation System (MRS), LOOPS is based on the Xerox Interlisp-D workstation.

The developers believe that design is an exploratory process in which specifications and goals evolve concurrently. Hence, the process is one of incremental refinement.

Palladio uses several design perspectives to specify a design. Perspectives are viewpoints from which the entire design or a part of it can be examined. Each perspective is composed of a set of primitive components and compositions which result from the application of composition rules to other components. A particular perspective allows the designer to see relevant details about a design while hiding others. The primitives and composition rules are selected such that errors will be revealed and avoided during the design process.

The CSG (Clocked Switches and Gates) perspective has as its primitives the steering switch, clocking switch, pull-up, pull-down, and controlled pull-up. The composition rules in this perspective indicate the permissible connections between primitives. For example, one rule specifies that a control input of a steering switch can only be connected to the output of a restoring logic gate. This rule reflects a particular correctness property, namely, that the level of a signal is reduced when it propagates through a pass trasistor and its level must be restored.

Palladio also allows the use of multiple perspectives simultaneously. This is useful when an engineer wants to design a subsystem in a top-down (depth first) manner. The subsystem can then be examined in detail while the rest of the system is "stubbed out" in some higher level perspective.

C-MU DAA

Thomas, Kowalski and others at Carnegie-Mellon University are exploring a new approach to the synthesis problem using a rule-based expert system [Kow83, Tom83b]. This system is called the Design Automation Assistant (DAA). DAA transforms a behavioral specification into control steps and data paths like RT/CAD. However, this translation is guided by a set of rules which model the kinds of decisions that real designers follow when decomposing a system into a functional block design. Preliminary results indicate that designs produced by DAA are of much better quality (fewer components and busses) than those produced by RT/CAD. With the addition of lower level representations and additional level to level, knowledge-based translators, a complete specification to implementation synthesis path will be formed.

USC Synthesis

Parker at the University of Southern California is investigating a formal method for the description and synthesis of digital systems [Haf83]. This method uses an algebraic description of behavior. The algebraic relations express the timing relationships which must be satisfied by any correct implementation of the hardware. Cost objectives are introduced to trade circuit delays against chip area. This method has been successfully applied to small example systems using CMOS/SOS standard cells, routed by the MP2D program.

Caltech

Ron Ayres has written an excellent book on silicon compilation that describes in detail the internal operation of an experimental silicon compiler [Ayr83]. As much as possible, layout is accomplished through cell abutment and stretchable geometry. Behavior is described in a symbolic register transfer language which is translated to a synchronous PLA-based controller. Structural interconnection is accomplished symbolically and is annotated by block diagrams that enhance designer (and reader) comprehension.

Case Western Reserve University

Several master's theses were initiated at Case to follow up specific aspects of the d system. An Adaptable Graph Editor (AGE), constructed by Michael Sorens, can be tailored for different graph modeling problems [Sor83]. This tool is based upon the observation that many graph models have similar characteristics (e.g., nodes interconnected by arcs, definition and instantiation of subnets, etc.) and that a common graph editing tool could be used for creating Petri Net graphs, block diagrams, or dataflow programs. With AGE, the user defines a set of primitive node objects from which arbitrary graphs can be drawn.

In his thesis, Chip Krauskopf used a graph schema to describe and synthesize self-timed nMOS circuits [Kra83]. The graph schema is more conventional than the d-n function notation and is similar to the Register Transfer Modules (RTM) of Bell [Bel72]. Although the translation process was fairly mechanical, the control circuitry was centralized, complicating the control to datapath wiring problem.

A symbolic structural description language, called TN, was developed by Bill Cook as part of his thesis work [Coo84]. TN supports hierarchical descriptions and a procedural notation for the instantiation and interconnection of components. (The procedural style makes the description of regular circuit arrays easy and concise.) A compiler for TN was constructed that selects and combines ISP behavioral models into flattened topology files for the N.mPc functional simulation system [Dro84, Ros84]. With minor modifications, the compiler can handle other kinds of behavioral modeling languages such as SPICE. Mixed mode topologies, where some components are modeled at the register transfer level and others at the logic level, can be generated by the TN compiler.

In the Logic Synthesis System (LSS) project funded by RCA, a compiler for a register transfer level behavioral language, called Wisp, is being developed. The compiler translates a Wisp description into a design database which uses an extended Petri Net graph schema for the representation of control information, and a datapath of large grain functional blocks (ALUs, register files, input and output pads, etc.). Using a technology dependent synthesis step, the design database is translated into a network of CMOS standard cells to be mechanically placed and routed with the MP2D physical layout program. The synthesis of functional blocks is guided by a library of block templates and will eventually be driven by a collection of "experts" to synthesize blocks from truth tables and logic equations. Although the input description language is different, many of the algorithms and database requirements of LSS are similar to those of the d system.

Appendix A

The d-n Notation

The description of d-n below is a terse, reference format presentation of the notation. Chapter 4 is a more pleasant introduction. Tool builders should be able to use this section to construct d-n-based tools. Details about the translation from the logical specification to its physical representation should help designers to employ d-n as a tool.

Tokens

The textual fields of a d-n description are composed of tokens. There are four kinds of tokens: <reserved-word>, <delimiter>, <identifier>, and <integer>. Tokens are built from simple characters according to these BNF rules:

<legal imbed> ::= . | _ | -
<sign> ::= + | - | <empty>
<identifier> ::= <letter> {<id char>}
<id char> ::= <letter> | <digit> | <legal imbed>
<delimiter> ::= . | , | :
<integer> ::= <sign> <digit> {<digit>}
<reserved word> ::= encapsulation | defined datatype |
e | dd | function | type |
Call | Controller | Entry | Exit

d-n Syntax and Informal Semantics

The description of d-n is presented in a series of tables and figures which depict the graphical symbols and textual constructs that make up a d-n specification. Information about the informal semantics of d-n and translation to circuitry are presented, also.

Encapsulation

Form: Encapsulation.
Constituents: Encapsulation structural diagram, encapsulation physical diagram.
Semantics: Encapsulations are high level, black boxes which describe the interior structure and floor plan of the chip design hierarchy.

Encapsulation Structural Diagram

Form: Encapsulation structural diagram.
Constituents: Instance symbol, interface point, connection arc.
Graphical syntax: See figure 58.
Syntax notes: The heading for this kind of diagram has the BNF syntax:
<heading> ::= encapsulation: <type name>
Lexical scope: The type name is global to the specification.
Semantics: The encapsulation structural diagram defines one or more instances of encapsulation and defined datatype subsystems and their interconnections. The connection arcs resolve the interconnection of external defined function calls to specific instances of those functions.

Figure 58. Encapsulation Structural Diagram

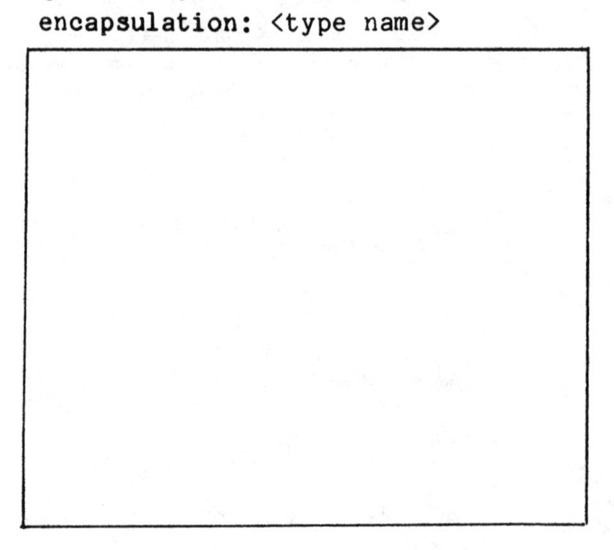

Instance Symbol

Form: Instance symbol.
Constituents: Interface point.
Graphical syntax: See figure 59.
Syntax notes: The label inside the instance symbol has the BNF syntax:
<instance label> ::=
<kind> : <instance name> type: <type name>
<instance name> ::= <identifier>
<type name> ::= <identifier>
Lexical scope: Since a type name refers to a subsystem specification, its scope is global. The scope of an instance name is that of the encapsulation structural and physical diagrams (i.e., the encapsulation itself).
Semantics: An instance symbol represents a real occurrence of a defined datatype or encapsulation.

Figure 59. Instance Symbol

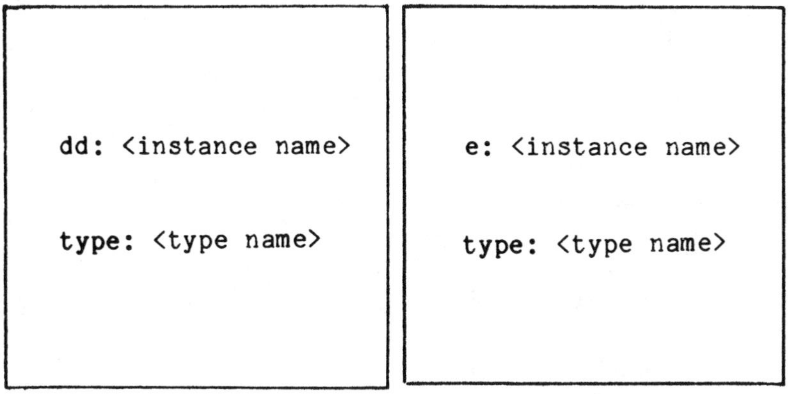

dd: <instance name> type: <type name>	e: <instance name> type: <type name>

Interface Point

Form: Interface point.
Constituents: Label (external interface only).
Graphical syntax: See figure 60.
Syntax notes: Interface points are drawn on the boundary of a structural diagram (external) or on the edge of an instance box (internal). Interfaces are labelled according to the BNF syntax:

<ST label> ::= <function name>
<function name> ::= <identifier>

Internal interface points belonging to an instance box are not labelled. They must follow the spatial arrangement of the external interface points contained in the specification of the instance type. That is, they must appear on the same edge of the box in the same order.
Lexical scope: In the case of an external interface point, the function name is meaningful (exported) outside the scope of the encapsulation or defined datatype.
Semantics: Each external interface represents either a function to be made available (arrow pointing out) for use at the interface or an external function to be called from within the enclosing subsystem (arrow pointing in). Interface points around an instance symbol represent logical connections to lower levels of the design hierarchy. An interface point that represents a receptor function is translated into a set of defined function inputs and outputs. An interface point representing an external function to be called is translated to a set of input (output) ports at the site of the call to that function. An interface point belonging to a defined function symbol represents the final linkage to the function inputs, outputs, and nodes. Interface points are also used to represent the local interfaces to arbitration and storage elements.
Implementation: Each interface point is translated to a set of physical connection points in some physical circuit cell.

Connection Arc

Form: Connection arc.
Constituents: Label (optional).
Graphical syntax: See figure 60.
Syntax notes: The label is optional and may be placed anywhere along the path of the arc.

Figure 60. Interface Points and Connection Arcs

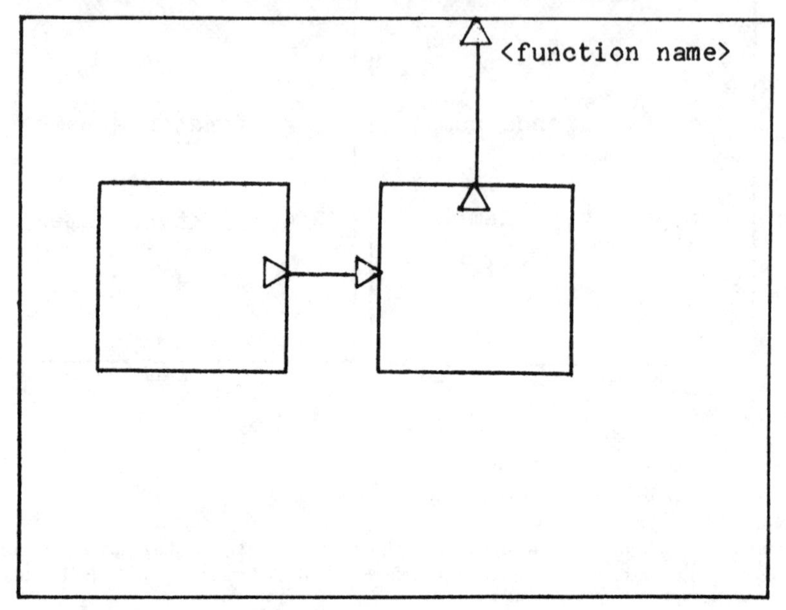

Semantics: A connection arc represents a communication bus between a call to a defined function and the actual instance of that function. It may also be used to connect a storage element (possibly through an intermediate arbitration element) and a primitive datatype function call. A connection arc from an external interface point to an internal interface either makes an internal function available for use or brings a set of connections to a function instance into the subsystem. If the external interface point represents a call (receptor) the internal interface point must be a call (receptor) at the next lower level of specification. If a connection arc links two internal interface points, one of the points must be a call and the other must be a receptor as defined by their respective type specifications. No more than one connection arc may be attached to any interface point. The label associated with a connection arc is purely descriptive.

Implementation: A connection arc is translated (by d-router) to a set of wires between the call and the function receptor instance.

Encapsulation Physical Diagram

Form: Encapsulation physical diagram.

Constituents: Encapsulation boundary, defined datatype boundary.

Graphical syntax: See figure 61.

Syntax notes: The BNF syntax for the diagram label is:

<e physical label> ::=

e: <type name> type: <type name>

<type name> ::= <identifier>

Lexical scope: The scope of the encapsulation type name is global.

Semantics: An encapsulation physical diagram allocates a physical region for the implementation of an encapsulation. It is a *scale* drawing with meaningful physical dimensions. The physical diagram shows the relative size and placement of each subsystem instance within the encapsulation. The subsystems are drawn and placed in accordance with the diagram scale factor, a quantity which is normally maintained by a d-editor. The encapsulation physical diagram is used to perform hierarchical space planning and design rule checks.

Implementation: The implementation of the encapsulation is defined by the individual implementations of its constituent subsystems.

Encapsulation Boundary

Form: Encapsulation boundary.

Graphical syntax: An encapsulation boundary is a labelled rectangle (fig. 61).

Syntax notes: The boundary label has the BNF syntax:

<e boundary label> ::=

e: <instance name> type: <type name>

<instance name> ::= <identifier>

An encapsulation boundary cannot cross any other boundary and cannot exceed the enclosing encapsulation boundary.

Lexical scope: The lexical scope of the instance name is the enclosing encapsulation. The type name is global.

Semantics: The rectangle is the same physical shape as the region that the encapsulation instance will occupy on the chip. The drawn dimensions of the rectangle are adjusted by the scale factor of the enclosing encapsulation physical diagram.

Figure 61. Encapsulation Physical Diagram

dd: <instance name> type: <type name>

e: <instance name> type: <type name>

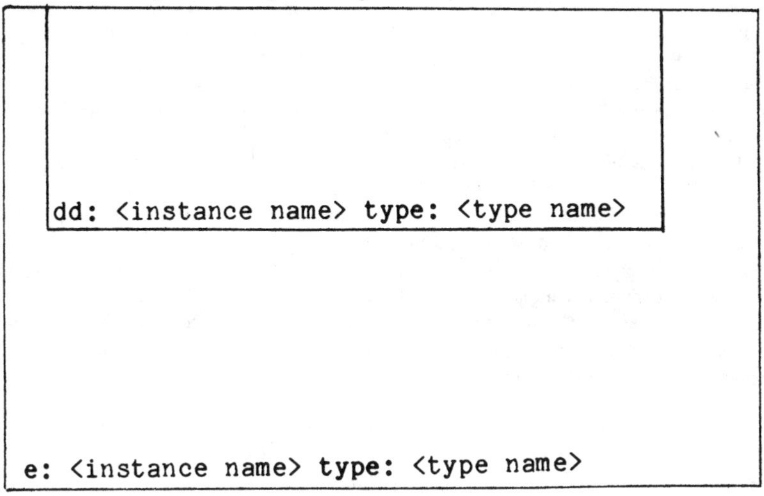

Defined Datatype Boundary

Form: Defined datatype boundary.
Graphical syntax: A defined datatype boundary is a labelled rectangle (fig. 61).
Syntax notes: The label has the BNF syntax:
<dd boundary label> ::=
dd: <instance name> type: <type name>
A defined datatype boundary cannot cross any other boundary and cannot exceed the enclosing encapsulation boundary.
Lexical scope: The scope of the instance name is limited to the enclosing encapsulation. The type name is global.
Semantics: The rectangle is the same physical shape as the region that the defined datatype will occupy on the chip. The drawn dimensions of the rectangle are adjusted by the scale factor of the enclosing encapsulation physical diagram.

Defined Datatype

Form: Defined datatype.
Constituents: Defined datatype structural diagram, defined datatype physical diagram.
Semantics: A defined datatype is the locus of system storage and functionality. It is the bottom structural level in the hierarchy.
Implementation: A defined datatype is translated to a set of real storage, arbitration, control, and computational circuits.

Defined Datatype Structural Diagram

Form: Defined datatype structural diagram.
Constituents: Store symbol, arbiter symbol, defined function symbol, interface point, connection arc.
Graphical syntax: See figure 62.
Syntax notes: The heading for a defined datatype (DD) structural diagram has the BNF syntax:
<DD heading> ::= defined datatype: <type name>
Lexical scope: The scope of the type name is global.
Semantics: The structural diagram defines the constituents of the defined datatype, namely storage, arbiters, and defined functions. The connection arcs indicate the interconnections between these elements and the defined functions.
Implementation: The arbiters and storage components are translated to real circuits through their d-tech-db templates. The implementation of a defined function is determined by its logical and physical specifications. Connection arcs are translated to wires by the d-router. Interior wiring may be replicated. Global wiring must be unique since it depends upon the absolute placement of the defined datatype on the chip.

Store Symbol

Form: Store symbol.
Constituents: Interface point.
Graphical syntax: See figure 63.
Syntax notes: The BNF syntax of the store symbol label is:
<store label> ::= <store name> <datatype name>
<store name> ::= <identifier>
<datatype name> ::= <identifier>

Figure 62. Defined Datatype Structural Diagram

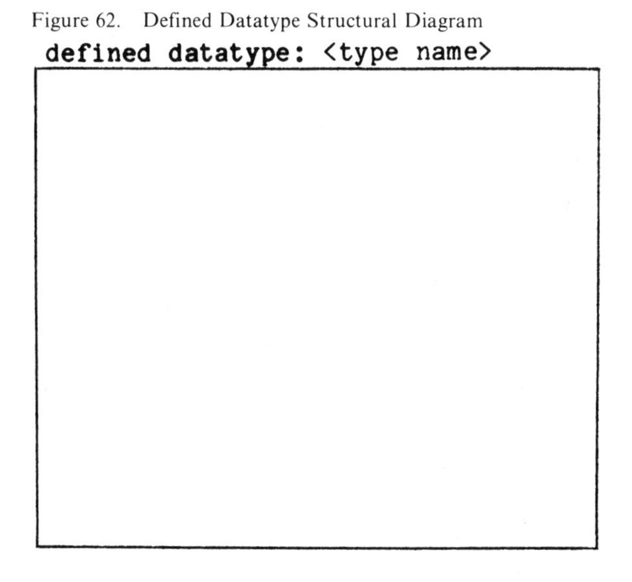

Figure 63. Store Symbol

<store name>

<datatype name>

Lexical scope: The scope of the store name is the defined datatype and its defined functions.

Semantics: A store symbol defines an instance of a primitive datatype—a datatype which can be implemented directly in hardware. The interface points around the store symbol represent the interfaces to the storage element called primitive datatype functions. Every primitive datatype has a set of these functions (as defined within the technology database) which may be used to manipulate an instance of the datatype and *only* that datatype. The actual semantics of a datatype (store) are determined by the d-tech-db templates for the datatype and its primitive functions.

Implementation: The datatype name selects a template in the d-tech-db which supplies the geometry for the datatype circuit. The designer must specify any instantiation parameters required by the template. The interface points are translated to connection points whose arrangement depends upon the circuit template.

Arbiter Symbol

Form: Arbiter symbol.
Constituents: Interface point.
Graphical syntax: See figure 64.
Syntax notes: The arbiter symbol label has the BNF syntax:
<arbiter label> ::= <arbiter name> <arbiter cell>
<arbiter name> ::= <identifier>
<arbiter cell> ::= <identifier>
Lexical scope: The arbiter name is local to the defined datatype and its defined functions.
Semantics: An arbiter controls shared access to a storage element. The actual arbitration scheme is defined by the d-tech-db template for the arbiter. The interface points around the arbiter symbol represent the interfaces through which access to a storage element is controlled.
Implementation: The arbiter cell name specifies a template in the d-tech-db which defines the implementation of the arbiter. The designer must specify any instantiation parameters required by the template. Each interface point is translated to a set of connection points whose arrangement around the cell is defined by the template, also.

Figure 64. Arbiter Symbol

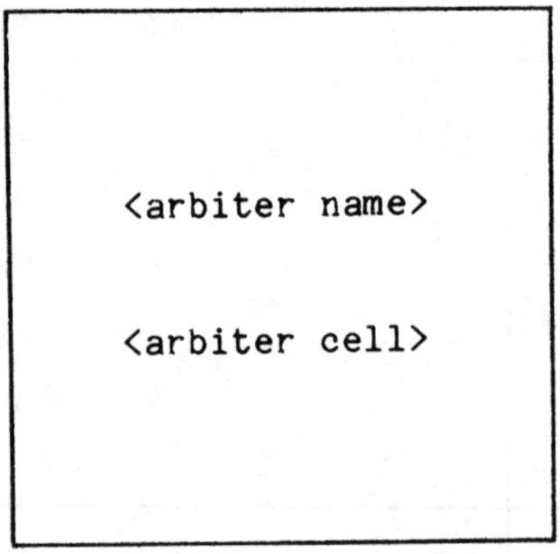

<arbiter name>

<arbiter cell>

Defined Function Symbol

Form: Defined function symbol.
Constituents: Interface point.
Graphical syntax: See figure 65.
Syntax notes: The BNF syntax for the symbol label is:
<fn symbol label> ::= <function name> function
Lexical scope: The scope of the function name is local to the enclosing defined datatype *unless* it is explicitly made available from the defined datatype via a connection arc connected to an external interface point. The scope of the name is then extended to the scope of the encapsulation which instantiates the defined datatype.

Figure 65. Defined Function Symbol

```
<function name>

function
```

Semantics: A defined function specifies some portion of the system behavior. The interface points around the function symbol represent the interfaces to defined function calls, primitive defined function calls, and the defined function inputs and outputs. There must be at least one interface point to represent the call interface to the function. For every defined function and primitive datatype function call in the specification of the function, there must be an interface point.

Defined Datatype Physical Diagram

Form: Defined datatype physical diagram.
Constituents: Store cell, arbiter cell, defined function boundary.
Graphical syntax: See figure 66.

Figure 66. Defined Database Physical Diagram

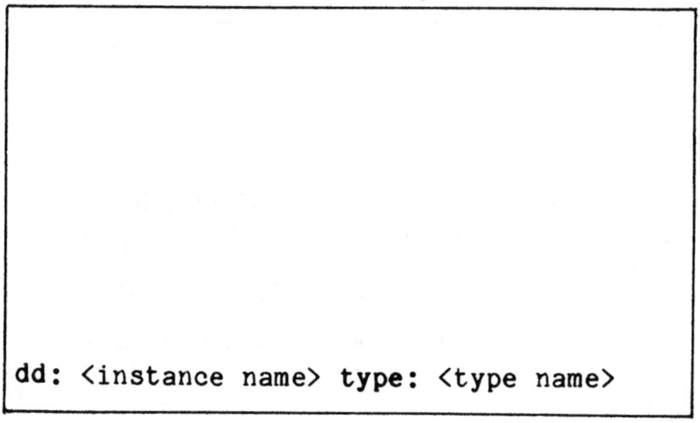

```
dd: <instance name> type: <type name>
```

Syntax notes: The BNF syntax for the DD physical diagram label is:
<DD physical label> ::=
dd: <type name> type: <type name>

Lexical scope: The type name is global.

Semantics: The DD physical diagram shows the relative size and placement of each constituent arbiter, store, and defined function. The diagram is a scale drawing with meaningful physical dimensions.

Store Cell

Form: Store cell.

Graphical syntax: See figure 67.

Syntax notes: See the BNF syntax for the store symbol label. The boundary of the store cell cannot cross any other boundary and cannot exceed the enclosing defined function boundary.

Semantics: The rectangular boundary is scaled according to the enclosing DD physical diagram. The actual size and shape of this boundary is determined by the d-tech-db template for the store datatype.

Implementation: The store cell represents a real nMOS circuit. Its geometry is defined by its d-tech-db template and instantiation parameters. These parameters must be specified by the designer. Any interface point on the corresponding store symbol in the DD structural diagram is translated to a set of connection points.

Figure 67. Store Cell

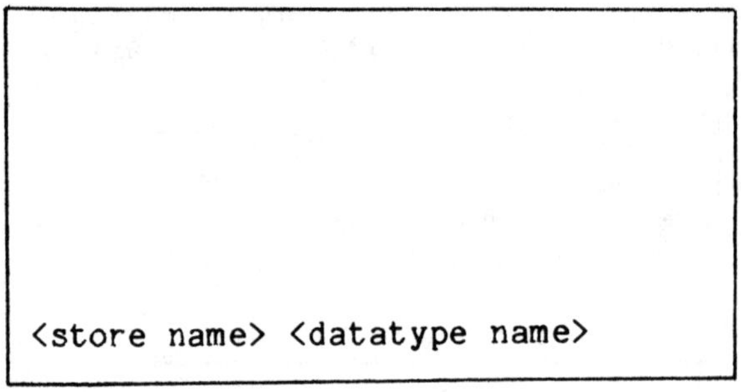

<store name> <datatype name>

Arbiter Cell

Form: Arbiter cell.

Graphical syntax: See figure 68.

Syntax notes: See the BNF syntax for the arbiter symbol. The boundary of an arbiter cell cannot cross any other boundary and cannot exceed the enclosing defined datatype boundary.

Semantics: The rectangular boundary is scaled according to the scale factor of the DD physical diagram. The actual size and shape of the boundary is determined by the d-tech-db template for the arbiter.

Implementation: The arbiter cell is a real nMOS circuit whose geometry is defined by its d-tech-db template. The designer must specify whatever instantiation parameters are required by the cell template. Any interface point associated with the corresponding arbiter symbol in the DD structural diagram is translated to a set of connection points.

Figure 68. Arbiter Cell

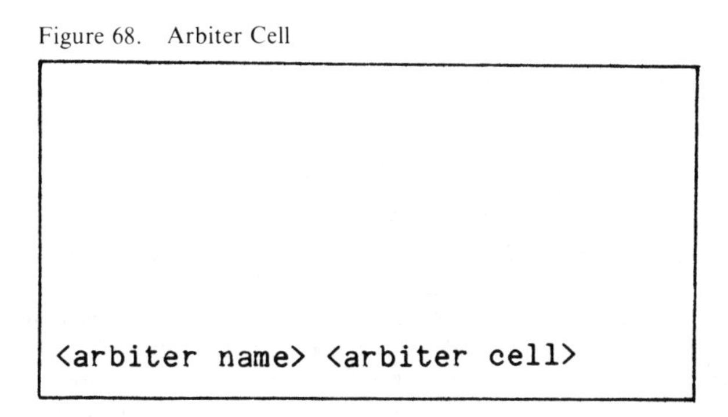

Defined Function Boundary

Form: Defined function boundary.
Graphical syntax: See figure 69.
Syntax notes: A function boundary label has the syntax:
<function boundary label> ::=
<function name> (function)
Lexical scope: The scope of the function name is the enclosing defined datatype.
Semantics: The defined function boundary is a rectangle drawn to the scale of the defined datatype physical diagram. It is the same shape as the physical region it will occupy on the chip.

Figure 69. Defined Function Boundary

<function name> (function)

Defined Function

Form: Defined function.
Constituents: Defined function logical diagram, defined function physical diagram.
Semantics: A defined function embodies some small portion of the system behavior.

Defined Function Logical Diagram

Form: Defined function logical diagram.
Consituents: Input symbol, output symbol, node symbol, arc symbol.
Graphical syntax: See figure 70.
Syntax notes: The heading has the BNF syntax:
<fn heading> ::= <function name> (function)
Lexical scope: Unless a function name is made available from a defined datatype, the scope of a function name is limited to its defined datatype.
Semantics: Arguments arrive at the defined function through its inputs. These data values flow along the arcs in the function graph and are processed by the nodes. The arcs determine the order in which intermediate results are consumed and produced—the data dependencies. Results are returned through the outputs.
Implementation: Arguments and results are binary encoded data values. The inputs buffer the argument values during function execution. The outputs hold the result values for the function caller. Some of the nodes are translated to real combinatorial data operations while others become states (and enabling signals) in the function controller. The data dependencies are translated into the transitions between controller states and data paths (wires) between combinatorial functions.

Figure 70. Defined Function Logical Diagram

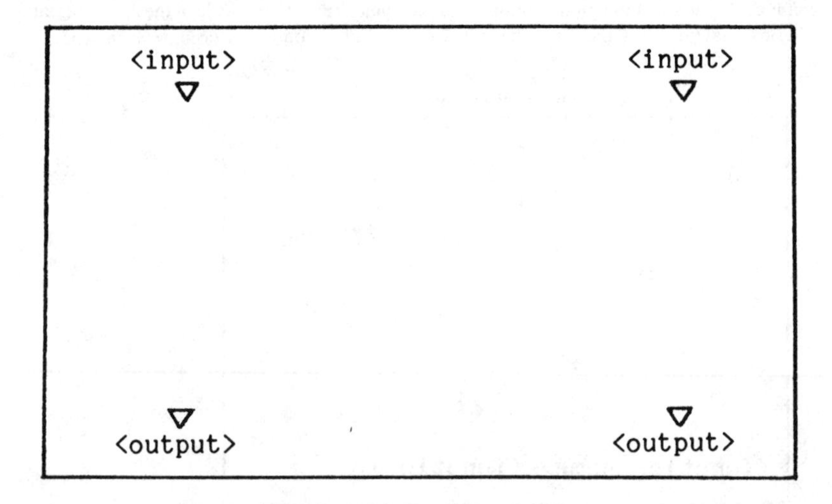

Input Symbol

Form: Input symbol.
Graphical syntax: See figure 71.
Syntax notes: The label on the input symbol has the BNF syntax:
<input name> ::= <identifier>
Lexical scope: The input name is local to the defined function.
Semantics: An input represents an incoming data value which is presented to the function when it is called.

Implementation: An input is translated to the d-tech-db template "Entry." Since arguments are represented as binary encoded values, the designer must specify the number of bits required to represent the incoming argument.

Figure 71. Input and Output Symbols

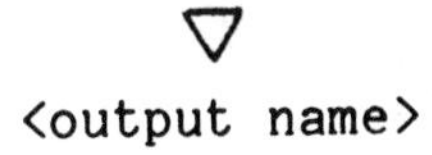

Input Cell

Form: Input cell.
Graphical syntax: See figure 72.
Syntax notes: The BNF syntax for the input cell label is:
<input cell label> ::= <input name> (Entry)
The boundary of an output cell cannot cross any other boundary and cannot exceed the boundary of the enclosing defined function physical diagram.
Semantics: The input cell is drawn as a scaled rectangle. The physical size and shape of the rectangle are determined by the template "Entry" and its instantiation parameters. The drawn dimensions of the rectangle are adjusted by the scale factor of the enclosing defined function physical diagram.
Implementation: The d-tech-db template "Entry" plus the designer-specified instantiation parameters define the geometry of the input cell. Currently, "Entry" is a static register and does not require a refresh clock. "Entry" has one control signal which enables the incoming data value into the register bits. This signal is asserted by the function controller during the "Entry" state.

Output Symbol

Form: Output symbol.
Graphical syntax: See figure 71.
Syntax notes: The label on the output symbol has the BNF syntax:
<output name> ::= <identifier>
Lexical scope: The output name is local to the defined function.
Semantics: An output holds a result computed by a function until the result has been returned to the caller.
Implementation: An output is translated to the d-tech-db template "Exit." Since a result is represented as a binary encoded value, the designer must specify the number of bits needed to represent the output value.

Figure 72. Input and Output Cells

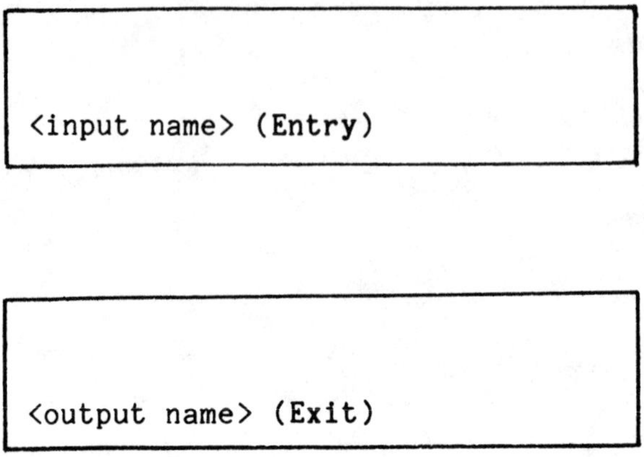

<input name> (Entry)

<output name> (Exit)

Output Cell

Form: Output cell.

Graphical syntax: See figure 72.

Syntax notes: The BNF syntax for the output cell label is:

<output cell label> ::= <output name> (Exit)

The boundary of an output cell cannot cross any other boundary and cannot exceed the boundary of the enclosing defined function and physical diagram.

Semantics: An output cell retains a result value computed by the defined function. It is represented by a scaled rectangle. The dimensions of the rectangle are determined by the d-tech-db template "Exit" and the instantiation parameters of the cell. The drawn size of the rectangle is adjusted by the scale factor of the defined function physical diagram.

Implementation: An output cell is translated to an instance of the d-tech-db template "Exit." The designer must specify the instantiation parameters for "Exit" to create the cell. "Exit" is a register which is wide enough to hold the binary encoded representation of the result. It has one control signal that enables the data value into the register bits. This signal is asserted by the function controller during the "Exit" state.

Node Symbol

Form: Node symbol.

Constituents: Primitive function symbol, primitive datatype function symbol, Call symbol, operator symbol.

Graphical syntax: Varies according to the kind of node.

Semantics: A node represents an information-processing function. The argument values for the operation are presented to the node through its input ports. The results produced by the operation are returned from the output ports. The number of input and output ports belonging to a node and their arrangement on the node symbol depend upon its d-tech-db template.

Implementation: The input ports and output ports translate to connection points within some real circuit cell. Other implementation characteristics depend upon the node template.

Port

Form: Port.

Graphical syntax: A port is drawn as a point on either the top edge (input) or bottom edge (output) of a node. An input (output) may not be connected to another input (output). More than one input may be connected to a single output.

Semantics: An input port is a data value sink. An output port is a data value source. The left to right ordering of ports on a given node is important. The ordering is used to match value sources with sinks by the d-router.

Implementation: Each port is translated to a set of connection points in a circuit cell. The number of arcs attached to an output port will affect the capacitive load driven by its physical counterpart.

Arc Symbol

Form: Arc symbol.

Graphical syntax: An arc is drawn as a (segmented) line between an output port and an input port, a function input and an input port, or an output port and a function output.

Semantics: The arc connects a data value producer to a data value consumer.

Implementation: An arc will be translated to a set of wires from the data source to the data sink. The sequencing implied by the arc will be used to create a control state transition that guarantees the production of the data value at the source before its consumption at the sink.

Primitive Function Symbol

Form: Primitive function symbol.

Constituents: Input port, output port.

Graphical syntax: See figure 73.

Syntax notes: The primitive function node is labelled with the name of the function:

<Fn name> ::= <identifier>

The number of input and output ports is determined by the technology database entry for the primitive function. Input ports appear on the top of the symbol box and output ports appear on the bottom.

Lexical scope: The primitive function name selects a template from the d-tech-db.

Semantics: The actual operation of a primitive function is determined by its d-tech-db template. Primitive functions are strictly combinatorial and do not possess history sensitive behavior.

Implementation: Every primitive function is translated to a real circuit as defined by its d-tech-db template and instantiation parameters.

Figure 73. Primitive Function Symbol

<Fn name>

Primitive Datatype Function Symbol

Form: Primitive datatype function symbol.
Constituents: Input port, output port.
Graphical syntax: See figure 74.
Syntax notes: The label in a primitive datatype function symbol has the BNF syntax:
<DTFn label> ::= <store name> . <DTFn name>
<DTFn name> ::= <identifier>
Lexical scope: The primitive datatype function name selects a d-tech-db template. The scope of the store name is the enclosing defined datatype.
Semantics: A primitive datatype function node invokes that function on a specific instance of a datatype. The datatype instance must be defined in the DD structural diagram. The primitive datatype of the function and the storage instance to which it is applied must be the same. The input (output) ports and the connections to those ports indicate the sources (sinks) for the arguments (results) of the function application.
Implementation: A primitive function node is translated to a controller state and a set of control wires. When the state is entered during execution, the signals invoke the desired function within the primitive datatype (storage) circuitry. The input and output ports belonging to the node are mapped to their sources and sinks (connection points). Since a primitive datatype function node does not generate any hardware other than a controller state, these connection points belong to other circuit cells. The graph arcs guide the d-router to the actual sources and sinks for the node.

Figure 74. Primitive Datatype Function Symbol

<store name>.<DTFn name>

Call Symbol

Form: Call symbol.
Graphical syntax: See figure 75.
Syntax notes: A call symbol label has the BNF syntax:
<Call label> ::= <Call name>
<Call name> ::= <identifier>
The Call name specifies the defined function to be called.
Lexical scope: The scope of the cell name is determined by the defined datatype structural diagram.
Semantics: The specified defined function will be called with the arguments provided by the input ports. Results will be returned from the output ports.
Implementation: A Call node is translated to a controller state that invokes the function call when it is entered. The assertion signal is sent to the receiving defined function controller which begins its execution by enabling the arguments into its "Entry" registers. When the called function enters its "Exit" state, it sends a completion signal back to the calling controller. The caller removes its invocation signal and, subsequently, the called function removes its completion signal. This implements an asynchronous, 4-cycle signaling protocol for function calls. The

input and output ports of a Call node are mapped to source and sink connection points—to connection points within other circuits in the defined function. The sources and sinks are identified by d-router according to the arcs in the function graph.

Figure 75. Call Symbol

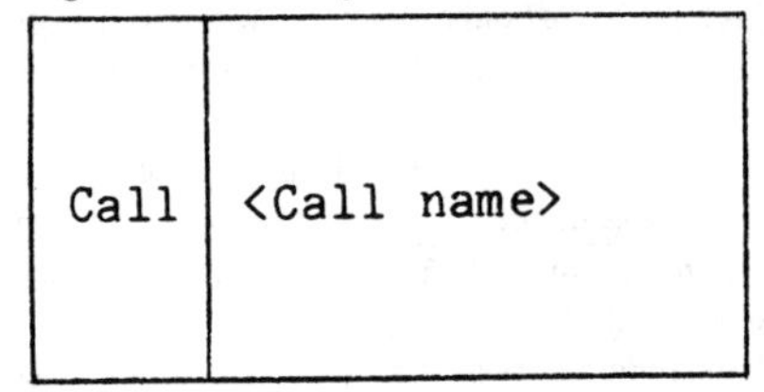

Operator Symbol

Form: Operator symbol.
Constituents: Call symbol.
Graphical syntax: See figure 76.
Syntax notes: The operator symbol label has the BNF syntax:
<Op symbol label> ::= <Op name>
<Op name> ::= <identifier>
Lexical scope: The operator name selects a d-tech-db template.
Semantics: The behavior of an operator depends upon its technology database template. Any control operation that requires a sequencing scheme other than a strict producer-consumer (data dependency) relationship must be implemented with an operator. Since cyclic function graphs are prohibited, iteration must be implemented using an operator.
Implementation: Operator nodes are translated to one or more controller states and condition sensitive transition logic. Using this logic, one or more defined functions can be called. Input and output ports are translated to value sources and sinks (connection points) which belong to other circuit cells in the defined function implementation. By tracing the arcs connected to the input and output ports, d-router can locate the actual sinks and sources.

Figure 76. Operator Symbol

<Op name>
Call <Call name>
o o o o o o
Call <Call name>

Primitive Function Cell

Form: Primitive function cell.
Graphical syntax: See figure 77.
Syntax notes: Every primitive function cell is labelled with the name of its primitive function. There must exist a cell template in the d-tech-db by this name. The boundary of a primitive function cell cannot cross any other boundary and cannot exceed the boundary of the enclosing defined function physical diagram.
Lexical scope: The lexical scope of a primitive function name is the enclosing defined function.
Semantics: A primitive function cell is represented as a scaled rectangle which has the same shape as its circuit geometry. The drawn dimensions of the rectangle are adjusted according to the scale factor of the enclosing defined function physical diagram.
Implementation: A primitive function cell is a real circuit whose geometry is defined by its d-tech-db template and the cell instantiation parameters specified by the designer. The input and output ports of the primitive function symbol are mapped to specific connection points arranged around the boundaries of the cell.

Figure 77. Primitive Function Cell

<Fn name>

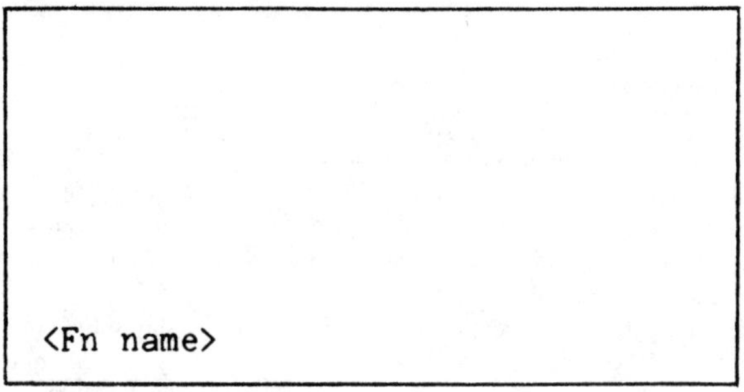

Controller Boundary

Form: Controller boundary.
Graphical syntax: See figure 78.
Syntax notes: The controller boundary has a label with the syntax:
<controller label> ::= <function name> (controller)
The controller (normally) has the same name as the defined function to which it belongs. The boundary of a controller cannot cross any other boundary and cannot exceed the boundary of the enclosing defined function physical diagram.
Semantics: The controller sequences the data operations to be performed by a defined function. It may be started asynchronously and keeps its own notion of absolute time. The controller is represented by a scaled rectangle. This rectangle has the same shape as the controller geometry and its drawn dimensions are adjusted by the scale factor of the defined function physical diagram.
Implementation: The function graph is translated to a set of states and transitions according to its data dependencies and the d-tech-db templates of its nodes. A Programmable Path Logic (PPL) circuit is generated which implements the equivalent state machine controller. Enabling signals are distributed as required. All controllers respond to the signal "Call" by starting execution and generate the completion signal "Exit." "Call" and "Exit" follow a 4-cycle signaling protocol. The combination of the signaling protocol and local clock implement a self-timed system.

Figure 78. Controller Boundary

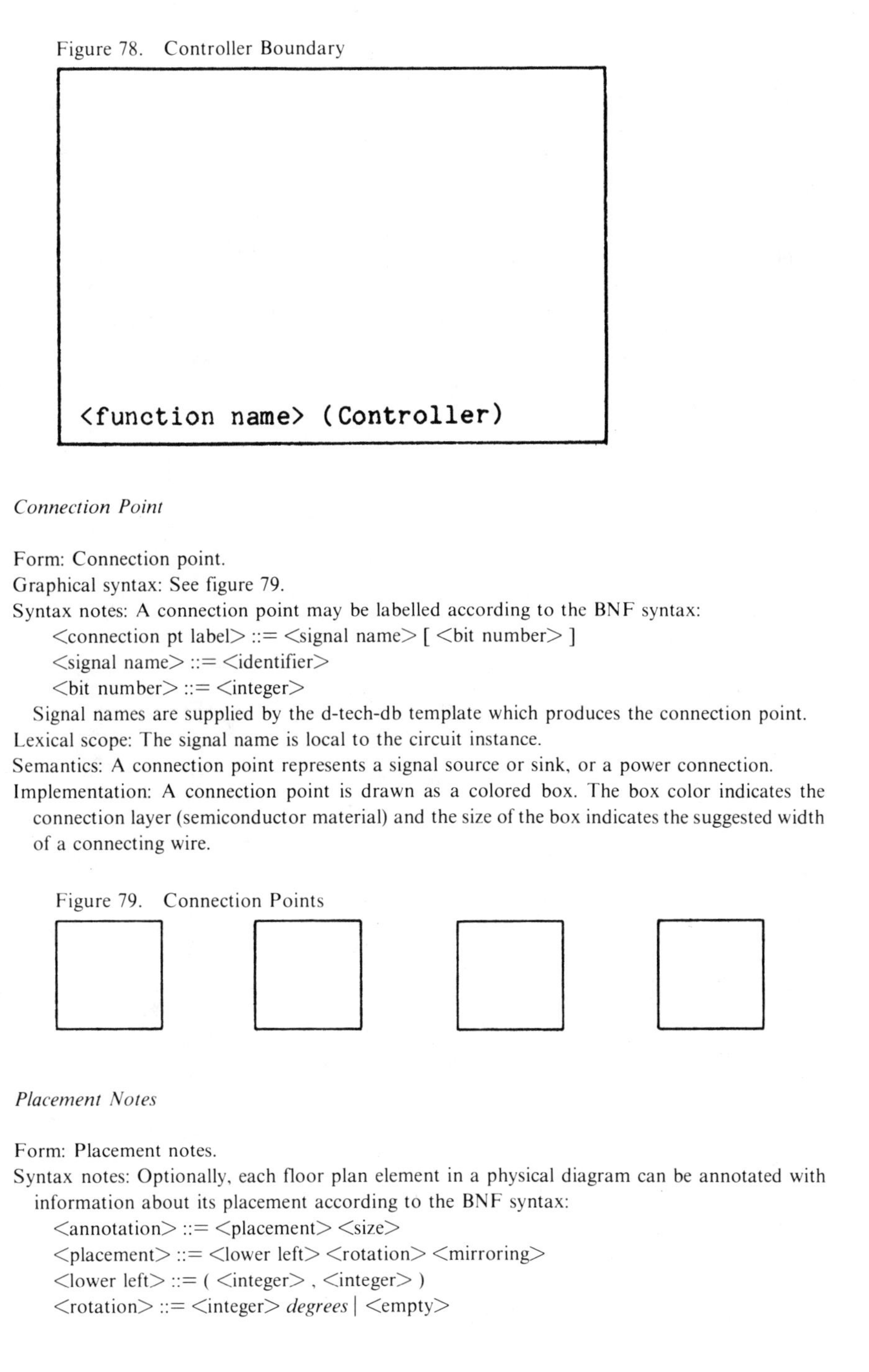

Connection Point

Form: Connection point.
Graphical syntax: See figure 79.
Syntax notes: A connection point may be labelled according to the BNF syntax:

<connection pt label> ::= <signal name> [<bit number>]
<signal name> ::= <identifier>
<bit number> ::= <integer>

Signal names are supplied by the d-tech-db template which produces the connection point.
Lexical scope: The signal name is local to the circuit instance.
Semantics: A connection point represents a signal source or sink, or a power connection.
Implementation: A connection point is drawn as a colored box. The box color indicates the connection layer (semiconductor material) and the size of the box indicates the suggested width of a connecting wire.

Figure 79. Connection Points

Placement Notes

Form: Placement notes.
Syntax notes: Optionally, each floor plan element in a physical diagram can be annotated with information about its placement according to the BNF syntax:

<annotation> ::= <placement> <size>
<placement> ::= <lower left> <rotation> <mirroring>
<lower left> ::= (<integer> , <integer>)
<rotation> ::= <integer> *degrees* | <empty>

<mirroring> ::= MX | MY | <empty>
<size> ::= (<integer> , <integer>)
Implementation: This information is used during translation to place circuits and subsystems.

Instantiation Parameters

Form: Instantiation parameters.
Syntax notes: Those d-n elements that are translated to d-tech-db circuit cells can be annotated (optionally) with their instantiation parameters according to the BNF syntax:
<I-params> ::= <integer> | <integer> , <I-params>
Semantics: The parameters will be used to instantiate the circuit template.

Appendix B

The Technology Database

d-n and its tools derive a great deal of their flexibility from the strength of the technology database, d-tech-db. This database contains a catalog of circuit cells which are as simple as an inverter and as complex as a content addressable memory.

The d-tech-db is based on a procedural database model; the entries in the database are procedures. When a procedure is executed, it produces an instance of a circuit cell or some information about the cell. By parameterizing the cell procedures, they become adaptable. For example, an inverter procedure may be parameterized for signal gain. The procedure describing the inverter geometry can then react to various values of gain to produce pull-up and pull-down transistors of sufficient size to achieve the specified gain.

This appendix presents the technology database model in terms of the procedural (template) model. It begins with a brief discussion of templates and cell instantiation. Then, the specific template operations (e.g., "produce geometry," "calculate delay," etc.) are described. Finally, a complete example template is presented.

Templates

A circuit cell is specified in a *template* which is a procedure with several *routines* that are just small executable programs. The act of executing a template is called *instantiation,* because the product of calling a cell template is an instance of the circuit cell. The templates are written in terms of *formal parameters* which are placeholders in the procedures for the actual values to be supplied when the template is called. These parameters are replaced by values, called *actual values,* which are provided by the caller during instantiation.

The partial template for an inverter is shown in figure 80. The template consists of a header describing the object type, formal parameters, cell inputs, cell outputs, and one routine with the selection label Space. When this routine is executed, it will return a pair of integers that describe the bounding box (maximum horizontal and vertical extent) around the circuit. Note that the size of the bounding box depends upon the magnitude of the parameter "Gain." (The syntax used in this section is merely suggestive since the development of a procedural language for cell specification is *not* a goal of this work. In the meantime, any modern programming language such as LISP or C can be used as a suitable replacement for a special template description language.)

Every template has the two formal parameters "Command" and "Name." The first parameter selects the routine to be executed during instantiation. Typically, there will be several different routines because the cell must be able to present several different views of itself to the database user. Each routine has a selector token. When the selector associated with a routine matches the actual value for "Command," that routine will be executed during the instantiation process. The formal parameter "Name" is a text string which may be used to annotate or identify the instance to be produced by the cell.

Figure 80. d-tech-db Template for an Inverter

```
Template       Inverter
Object type    Primitive function
Parameters     Command, Name, Gain.

Inputs  Fixed at 1. Bit
Outputs Fixed at 1. Bit-complement
Control None.

Routine Space

      if Gain leq 4 then return < 16, 34 >
      else return < 16, 26 + Gain * 2 > fi
```

Since there may be many different views of a given cell, there will be several different selectors and routines. Some possible selectors and their uses are:

—Attributes: Return the static information (the header) about a cell such as the number and names of the inputs and outputs and the kind of d-n operation that it implements.

—Geometry: Produce the detailed circuit geometry for a cell.

—Space: Return a pair of numbers indicating the maximum horizontal and vertical extent of the cell (i.e. its bounding box).

—Speed: Calculate and return the worst case delay through the circuit.

—Connection: Return a list structure containing the connection points for an instance. This is used for wire list generation.

—Control: Return information regarding cell control such as enabling signals or machine state requirements.

—LogicalSimulation: Output a gate or register transfer level description of the circuit suitable for use in a logic simulator.

—CircuitSimulation: Generate a transistor level description of the cell to simulate its electrical behavior.

The procedural scheme does not constrain the imagination of the cell designer. Anything which can be expressed as a program is a candidate for the cell database.

Template Routines

The following subsections describe some of the template routines and the operations that they perform.

The Template Header

The template header assigns a name to the template definition, declares its object type, and specifies its formal instantiation parameters (fig. 81). The template name acts as a database key; it is used to

Figure 81. Template Header

```
Template       Template name
Object type    d-n object type
Parameters     Command, Name, Parameter names.

Inputs         Fixed or Variable. Input names.
Outputs        Fixed or Variable. Output names.
```

select a template for instantiation. Therefore, template names must be unique. The template object type must be selected from the set:

—Primitive function.

—Primitive datatype.

—Primitive datatype function.

—Operator.

—Special.

This descriptor specifies the kind of d-n function performed by the cell. It is used to form the semantic context surrounding an instance of the cell within a d-n design. The parameter list declares the formal instantiation parameters for the template and it must include "Command" and "Name."

Next, the template contains a list of the logical inputs and outputs belonging to the cell. In a defined function logical diagram, the inputs and outputs are the ports around a node symbol. In the physical domain, each input or output is translated to one or more signal sinks or sources called *connection points.* The number of inputs and outputs *per cell* may be fixed or variable. When the number is fixed, the cell template cannot adapt for a variable number of I/O connections as in the case of the inverter. Some circuits, such as NAND and NOR, are expandable in regular, mechanical ways (fig. 82). Hence, a NAND gate could be parameterized for a variable number of inputs. When a variable number of inputs or outputs are permitted, a minimum number may be specified. For example, a NAND gate would require at least two inputs since a one-input NAND gate is somewhat meaningless.

Figure 82. Header for a Variable Geometry NAND Gate

```
Template      Nand
Object type   Primitive function.
Parameters    Command, Name, Inputs.

Inputs   Variable 2 minimum.
Outputs  Fixed at 1.
```

These input and output definitions are used to form the conceptual linkage between the logical specification of a cell instance and its physical one. If a logical port is drawn on a graph node, there must be a corresponding set of physical connection points in the implementation of that node. The d-editor uses this header information about the inputs and outputs to create the ports associated with a node. Later, when the physical translation of a design is in progress, the match is made between logical ports and physical points.

Geometry

When a routine selected by Geometry is executed, it will produce the physical cell geometry. The transistors and wires that make up the circuitry are described in terms of the geometric primitives Layer, Box, and Pad (fig. 83). Material is specified by a Layer statement. For example, the statement "Layer Poly" means that polysilicon will be used to make any of the boxes following it until the next Layer statement when some other material will be selected. Boxes are specified by the coordinates of the lower left and upper right corners of the box. A pad is a square box and is defined by the coordinate of its lower left corner and the length of a side.

By convention, the origin of an instance is the lower left corner of its geometry. Hence, all coordinate (X,Y) values in a geometric primitive such as Box must be greater than or equal to zero. By standardizing the relative position of an instance with respect to its origin, some advantages are gained:

—Placement: Translation is straightforward. Rotation is performed around the lower left corner.

—Representation of boundary size: Can be stored in 2 integers.

—Design rule checks: With conservative layout, design rule checking can be limited to the boundary of the cell. Everything inside the boundary can be assumed to be correct.

Figure 83. Geometric Primitives

```
Layer material
Box   lower-left-corner upper-right-corner
Pad   lower-left-corner length-of-a-side
Instantiate template at lower-left-corner
    with parameters.
Translate to coordinates
Rotate by number degrees
Mirror around X or Y
Revert
```

Besides the position of the origin, the unit of distance must be established, too. Each unit is equal to one-half the minimum feature size of the circuit technology. Circuits are arranged on a one-half lambda grid where lambda is the minimum feature size. By using a virtual grid unit instead of an absolute distance, cell designs can be scaled down if conservative design rules are employed during cell layout [Lyo81]. Fractional units are not permitted so all coordinate values must be integers.

Space

A routine selected by Space will return a pair of values indicating the maximum vertical (Y-axis) and horizontal (X-axis) extent of the instance. (Remember that the lower left corner of the instance is the origin.) These values can be used for auto-scaling, simple design rule checks, etc.

Figure 84. Speed Routine for the "Inverter"

```
Template    Inverter
Object type Primitive function
Parameters  Command, Name, Gain, Load.

Speed
    return 14.0 * (Load / 0.05E-12)
```

Speed

The interface between the technology database and the delay analyzer, d-da, is the routine selected by Speed. The Speed routine returns the worst case delay through the instantiated circuit. In figure 84 the estimated delay depends upon the actual value of "Load," the capacitive load (picofarads) to be driven into the output. The time value returned is in nanoseconds. This equation is a simplistic estimate of the inverter delay since the actual delay also depends upon the length of the pull-up transistor—a quantity that depends upon the parameter "Gain." One advantage of the procedural database style is its ability to incorporate better models for circuit behavior once those models have been determined through simulation or experimentation. Further, since delay is not represented by a static quantity, it can be calculated dynamically depending upon the electrical context of a cell.

During delay estimation, a template can prompt the designer for additional information about the design or how the cell is to be used. In the case of the operator "Conditional," which conditionally calls a defined function, the delay when the incoming condition signal is "true" will be very different from the "false" branch delay. Figure 85 contains the Speed routine from "Conditional." It will evaluate the delay of the "true" or "false" branch of the "Conditional" depending upon the response of the designer.

Figure 85. Speed Routine for "Conditional"

```
Template    Conditional
Object type Operator
Parameters  Command, Name, Function.

Inputs  Variable 1 minimum Condition.
Outputs       Variable 1 minimum.

Routine Speed
    Print("Information request from Conditional ", Name)
    Print("Analyze true or false branch? [t f] ")
    Branch := Read()
    if Branch = 'f' then return 30.0
    else return delay of Function f1
```

Lumping circuit delay into a single value is somewhat artificial. In circuits possessing more than one input or output (such as "Oneof2" in the next section), there will be more than one electrical path through a circuit with its own delay. Use of the worst case, however, guarantees a conservative design and increases the probability that the real circuit will function correctly.

The equation that is used to find the delay is derived through electrical simulation of the circuit under varying amounts of capacitive load. The circuits employed in the cache memory example of chapter 6 were simulated to determine their electrical delays. The assumptions made about incoming signals, geometry, and material properties are summarized in table 6.

Control

The Control routine returns information about the control requirements of a cell instance. For example, the state table generator, d-FSM, must know if a cell needs a machine state to enable its operation. At the present time, this is the only information returned by a Control routine.

Table 6. Assumptions on the Electrical Simulation of Cells

Incoming signals		
Rise time	10 nsec	
Fall time	10 nsec	
Low signal	0 volts (Infinite compliance)	
High signal	5 volts (Infinite compliance)	
Feature size	2 microns	
Capacitances		
Gate	1.OE-15 F/square-micron	
Polysilicon	1.OE-16 F/square-micron	
Diffusion	2.5E-16 F/square-micron	
Metal	0.75E-16F/square-micron	
Resistances		
Gate	1.OE4	ohms/square-micron
Polysilicon	150	ohms/square-micron
Diffusion	25	ohms/square-micron
Metal	0.075	ohms/square-micron

Connection

A Connection routine returns a list structure which describes the geometric positions of the connection points embedded in the cell instance. Figure 86 illustrates the overall structure of a connection list. The structure consists of four sublists: one sublist for the inputs, outputs, control, and power connections. Within each sublist are more sublists, each describing a particular input (or output, etc.) with the following information:

—Connection name: A name identifying which particular input, output, etc. this list describes.

—Material: The material layer on which the connection points will be placed.

—Size: Physical width of each connection point box.

—Point list: List of coordinate pairs each defining the position of a particular connection point. Ordered from bit 0 to bit N.

For those cells with a parameterized number of inputs, the number of input sublists will vary according to the actual parameter value (fig. 87). Since there is a list of connections for each input or output and there is a port in the logical diagram for each physical one, there will be a connection list for each logical port.

Figure 86. Connection List Structure

```
<
<Inputs
    <
    <Input-name Material Size < Point-list >>
    >>
<Outputs
    <
    <Output-name Material Size < Point-list >>
    >>
<Control
    <
    <Control-name Material Size < Point-list >>
    >>
<Power
    <
    <Vdd    Metal Size < Point-list >>
    <Ground Metal Size < Point-list >>
    >>
>
```

Figure 87. Connection Routine for the Inverter Template

```
Routine Connection

    Height := if Gain<=4 then 56 else 2*Gain+48 fi

    Return
        <
        <Inputs,
            <<Input, Poly, 4
                <for i=0:Width-1 <0, i * Width + 15>>>>>
        <Outputs,
            <<Output, Poly, 4
                <for i=0:Width-1 <36,i * Width + 27>>>>>
        <Power,
            <<Vdd,    Metal, 8
                <<28, 0> <28, Height * Width - 8>>>
             <Ground, Metal, 8
                <<4,  0> <4,  Height * Width - 8>>>>>
        >
```

Template Example

The example presented in this section is a 2-to-1 data selection (multiplexer) circuit. The schematic for the selector is shown in figure 88. This circuit has four inputs ("SelectA," "SelectB," "InputA," and "InputB") and one output, "Output." When either "SelectA" or "SelectB" are raised to Vdd (logical one), the data signal "InputA" or "InputB" will be selected (respectively) and passed to the buffer stage. It is assumed that only one of the select lines will ever be raised to Vdd, while the other line is pulled to ground (logical zero).

Figure 88. Schematic for a "Oneof2" Selector

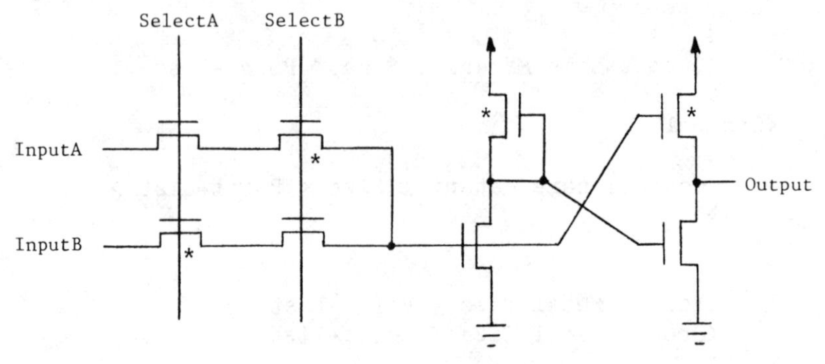

The physical device consists of a selection matrix of pass and high implantation density depletion transistors and a buffer element. The layout for the selector is shown in figure 89.

Figure 89. Circuit Geometry for "Oneof2"

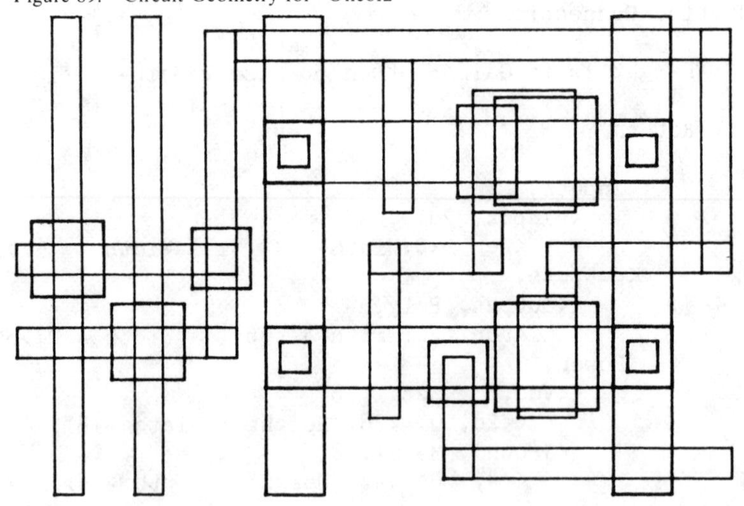

Figures 90 and 91 contain the template for a 2-to-1 selection circuit. This cell is called "Oneof2," an identifier reflecting its function. Any reference to the cell is made through this identifier. The template has four formal parameters: "Command," "Name," "Width," and "Load." The first two parameters are mandatory as mentioned earlier. The parameter "Width" specifies the number of individual selection circuits to generate in the complete instance. It is the number of bits an instance of the circuit will be able to switch. "Load" is the estimated capacitive load on the output of "Oneof2." The header declares the four inputs, "SelectA," "SelectB," "InputA" and "InputB," and declares the output, "Output." The number of inputs and outputs is fixed; additional inputs and outputs cannot be defined.

Figure 90. Template Header and Geometry for "Oneof2"

```
Template      Oneof2
Object type   Primitive function
Parameters    Command, Name, Width, Load.

Inputs  Fixed at 4. SelectA, SelectB, InputA, InputB.
Outputs Fixed at 1. Output.
Control None.

Routine Geometry
    Loop for i:=0:width translate to (0,i*63).

    Instantiate Buffer at (0,30)
    with Geometry,Name,1.

    Layer Diffusion
        Box (0,29)  (30,33)
        Box (0,18)  (30,22)
        Box (26,18) (30,33)
    Layer Poly
        Box (5,0)   (9,63)
        Box (16,0)  (20,63)
        Box (26,29) (30,61)
    Layer Implant
        Box (2,26)  (12,36)
        Box (13,15) (23,25)
    Layer buried
        Box (24,27) (32,35)
    Revert to previous spatial environment.
    EndLoop
```

There are five routines in "Oneof2" with the selectors Geometry, Space, Speed, Control, and Connection.

Figure 91. Template for "Oneof2" Continued

```
Routine Space

    return < 97, 63 * Width >

Routine Speed

    return 5.0 + (10.0 * (0.2E-12 / Load))

Routine Control return False

Routine Connection

    return
        <
        <Inputs
            <
            <InputA,  Poly, 4,
                <for i=0:Width-1 <0, i * Width + 29>>>
            <SelectA, Poly, 4,
                <<16, 0> <16, 63 * Width - 4>>>
            <SelectB, Poly, 4,
                <<5,  0> <5,  63 * Width - 4>>>
            <InputB,  Poly, 4,
                <for i=0:Width-1 <0, i * Width + 18>>>
            >>
        <Outputs
            <
            <Output, Poly, 4,
                <for i=0:Width-1 <93, i * Width + 2>>>
            >>
        <Power
            <
         <Vdd,    Metal, 8,
                <<81, 0> <81, 63 * Width - 8>>>
         <Ground, Metal, 8,
                <<34, 0> <34, 63 * Width - 8>>>
           .>>
        >
```

In the Geometry routine, the loop surrounding the Layer and Box statements is controlled by the parameter "Width." When executed, the loop will produce several copies of the selector circuit. The number of copies is determined by the actual value supplied for "Width" at instantiation time. The selection matrix is generated with Layer and Box primitives. The buffer circuit is produced through another level of cell instantiation. The geometric information accompanying the Instantiate statement will place the buffer relative to the origin of the selector cell. A cell can be rotated or mirrored, but in this case, neither rotation or mirroring is requested. Through careful foresight, the output of the selection matrix connects with the input of the buffer element.

The Space routine returns a pair of integers that describe the overall dimensions, the bounding box, of the instantiated selector. Since the size of the bounding box will vary with the number of individual selection circuits to be generated, the Space routine must be parameterized by "Width." The Control clause returns False, indicating that "Oneof2" does not require a machine state to enable its operation. The Speed routine uses "Load" to calculate and return the maximum signal delay through "Oneof2." Finally, the Connection routine generates and returns a connection list which describes the connection points within the circuit instance.

Appendix C

Control

This appendix discusses the translation from a function graph specification to the state machine which sequences its execution. The target language for translation is a state table showing the states that the controller can assume and the logic selecting those states. The table can be translated to a "one-hot" state machine (and its equivalent circuitry) using the tools and techniques of Carter and Hollaar [HolIE, Car81].

Discussion

A state diagram is a directed graph. It describes the behavior of a state machine which is the execution controller for a particular defined function. The places or nodes in the graph represent the states which the machine can assume. The directed arcs show the transitions between the states. The arcs are labelled with the enabling conditions that control the transfer from one state to another. The same information can be kept in a state table which is somewhat harder to interpret visually, but is more conducive to mechanical generation and analysis. State diagrams will be used to show interesting control topologies while state tables will be used as the target representation for control information.

Figure 92 contains a sample state diagram for a d-n controller. The equivalent state table for this diagram is shown in figure 93. There are three "special" states:

1. "Initial" is the place where execution begins at system start-up. It is also the place where the controller waits for a function call.
2. "Entry" is entered after a function call is asserted through the enabling condition "Call."
3. "Exit" is active during the function return sequence. It leads back to Initial and is enabled by the condition "not Call."

All state tables and diagrams derived from a d-n defined function will have these three states and transitions. They implement the control portion of the function call mechanism.

The example in figures 92 and 93 demonstrates concurrent path execution. States 3 and 4 will eventually be started as a consequence of an active state 2. The lower execution path will pass through state 5. Both of these paths end at state 6. Execution must be synchronized before starting state 6 because it may take a longer time to sequence through the lower path. Synchronization is specified through the enabling condition "Synch 3,5." The list of states following the keyword "Synch" indicates all the states which must be active before the transition is made.

The transition from "Initial" to "Entry" is enabled by the condition "Call." This condition is asserted by the caller to start execution of the defined function. The transition from "Exit" to

Figure 92. Example State Diagram

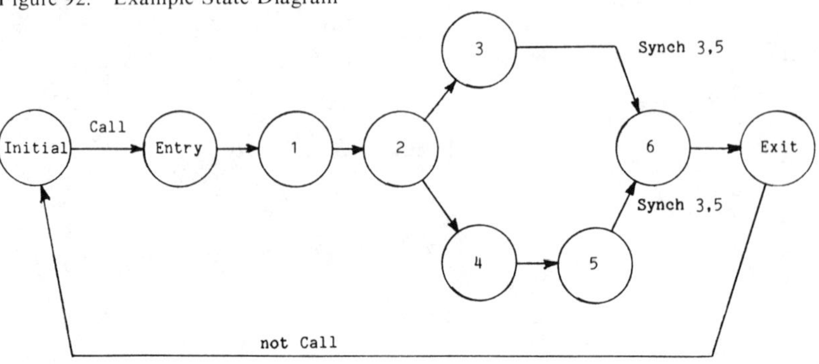

Figure 93. Equivalent State Table

State ->	Next state	: Enabling condition
Initial	-> Entry	: Call
Entry	-> 1	
1	-> 2	
2	-> 3	
2	-> 4	
3	-> 6	: Synch 3,5
4	-> 5	
5	-> 6	: Synch 3,5
6	-> Exit	
Exit	-> Initial	

"Initial" is enabled by the condition "not Call." The signals "Call" and "Exit" correspond to request and acknowledgement rails and implement a four-cycle asynchronous handshake (fig. 94). The caller asserts "Call" to request the function execution and removes "Call" in response to "Exit." "Exit" will be removed as the transition from the "Exit" state to "Initial" is made. One other enabling condition is required to complete the function call control linkage. Since the caller suspends execution until an acknowledgment signal is received, the enabling condition "Ack" is required on the transitions leading from the state associated with a function call.

Enabling conditions can be used to introduce conditional execution into a graph. Figure 95 contains a diagram illustrating an "if . . . then . . . else" type of conditional execution. Note that the diagram topology is similar to that of concurrent execution, but the separate execution paths are not synchronized. Nondeterministic behavior is avoided since the two paths are mutually exclusive—only one path will be active.

Since assumptions about mutually exclusive path execution may induce conceptual errors at the specification level, conditional execution is ruled out in function graphs (fig. 96). Every arc in a function graph will eventually be active during any execution of the function. Conditional operation as well as iteration must be specified through control operators. Hence, the use of conditional state transitions is reserved for the implementation of control operators, presumably a constrained, well-designed situation.

Figure 94. 4-Phase Function Call Handshake

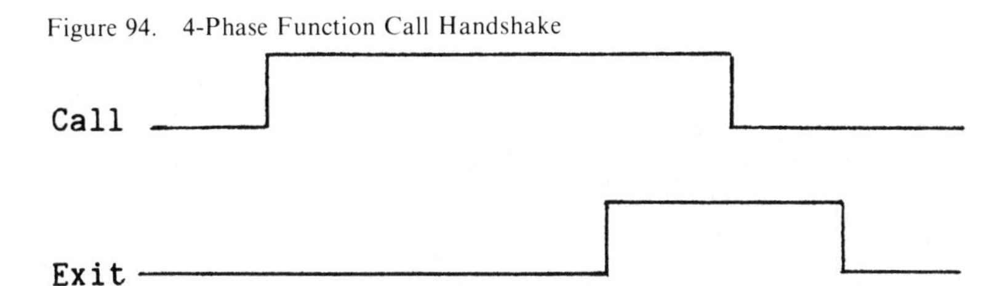

Figure 95. Conditional Execution

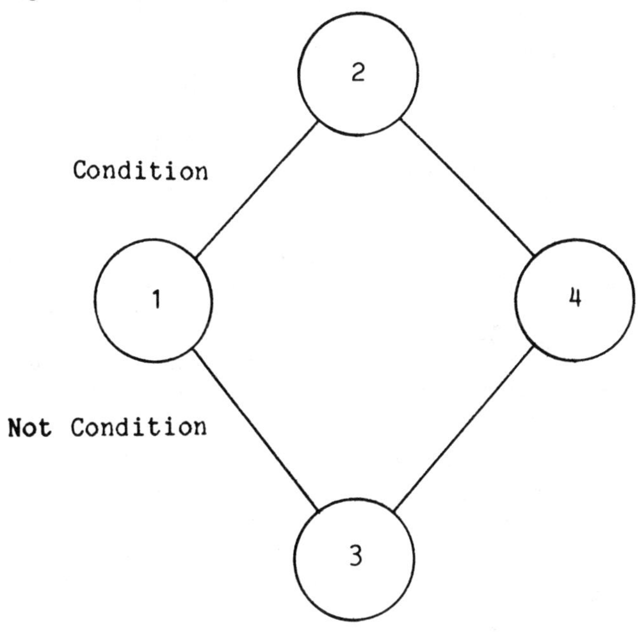

Figure 96. State Table for Conditional Execution

State	->	Next state	:	Enabling condition
1	->	2	:	Condition
1	->	3	:	not Condition
2	->	4		
3	->	4		

The translation from a defined function specification to its associated state table is performed in the following way. The function graph is rewritten with two special nodes "Entry" and "Exit." The function inputs are rewritten as the output ports on "Entry" and the outputs become the input ports on the node "Exit." The two special states "Entry" and "Exit" are assigned to the nodes "Entry" and "Exit," respectively (fig. 97).

Figure 97. Special Nodes "Entry" and "Exit"

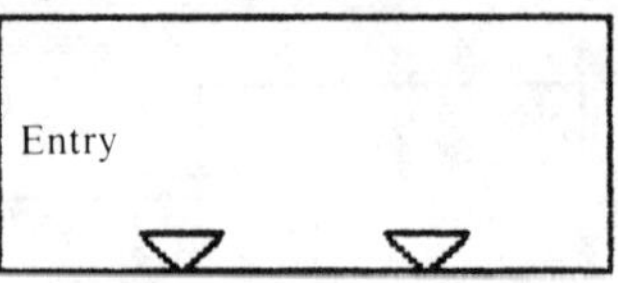

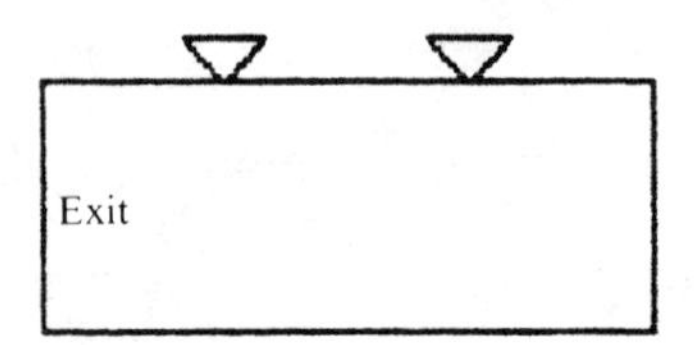

Next, each of the nodes in the rewritten graph which requires a machine state for its execution are marked. The technology database supplies the necessary information to complete this step. The nodes "Entry" and "Exit" are always marked.

The algorithm uses three lists. One of these lists, NL, is a list of graph nodes and is initialized with the special node "Exit." SL is a list of states and it is initialized with the states "Initial," "Entry," and "Exit." When the algorithm terminates, SL will contain all the states in the controller. TL is a list of state transitions which is initialized with the transitions:

Initial → Entry : Call
Exit → Initial.

These transitions provide for the function entry and exit sequence. The enabling condition "Call" will start execution upon a call to the function.

The algorithm operates by working back from the node "Exit" by finding the marked nodes which precede it temporally. A machine state is created for each marked node encountered. A transition from each of these states to "Exit" is recorded. If more than one transition leads to "Exit," the synchronizing condition "Synch" is added to each of the transitions with the appropriate state list. "Exit" is removed from NL and is replaced by the marked nodes which are its predecessors. Processing is applied to the nodes in NL until the only node remaining is "Entry," which has no predecessor. Unmarked nodes are ignored except as a source for additional precedence (dependency) relationships.

Figure 98 illustrates the first step in the control translation process—rewriting the function graph with "Entry" and "Exit" nodes. The nodes requiring a machine state have been marked with a "*" character. Initially, the three working lists contain:

NL: "Exit"
SL: Initial, Entry, Exit

TL: Initial → Entry : Call
Exit → Initial.

Figure 98. "Fetch" Rewritten with "Entry" and "Exit"

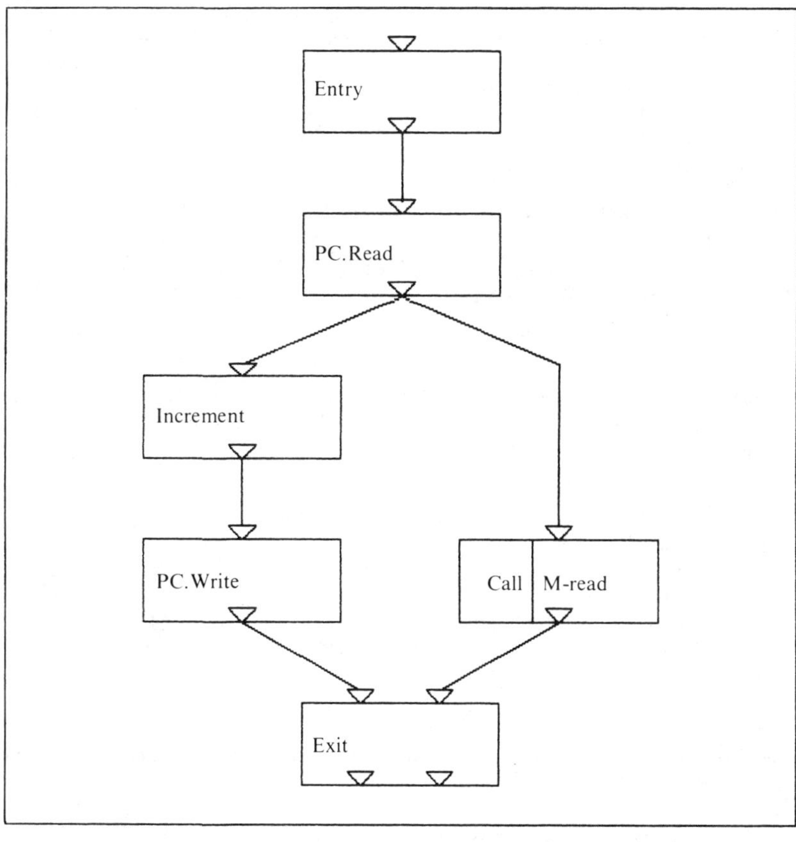

Starting with "Exit," the marked nodes which precede it are "PC.Write" and the call to "Mem-read." Two states, 1 and 2, are created for these nodes and are added to SL. Transitions from 1 and 2 to "Exit" are added to TL. They must have the synchronizing condition "Synch 1,2" because 1 and 2 both lead to "Exit." Since state 2 is associated with the Call node "Call Mem-read," the Ack condition must also be specified so that the execution of that path can be restarted asynchronously after "Mem-read" executes and returns. "Exit" is replaced in NL by "PC.Write" and "Call Mem-read." These steps leave the working lists with the contents:

NL: "PC.Write," "Call Mem-read"
SL: Initial, Entry, Exit, 1, 2

TL: Initial → Entry : Call
Exit → Initial
1 → Exit : Synch 1,2
2 → Exit : Synch 1,2 Ack.

During the next step, the paths from "PC.Write" and "Call Mem-read" are traced back to their predecessors. On the path leading back from "PC.Write," the "Increment" node is unmarked, so it is used simply as a source for additional dependency relationships. Both paths lead to "PC.Read." So, the nodes in NL are replaced by "PC.Read," one new state (3) is added to SL, and two new transitions are put into TL. The lists now contain:

NL: "PC.Read"
SL: Initial, Entry, Exit, 1, 2, 3

TL: Initial → Entry : Call
Exit → Initial
1 → Exit : Synch 1,2
2 → Exit : Synch 1,2 Ack
3 → 1
3 → 2.

The predecessor of "PC.Read" is "Entry." A transition from Entry to 3 is added to TL and "PC.Read" is replaced in NL by "Entry." Since "Entry" is the only node in NL, the process stops here, leaving the following states and transitions:

SL: Initial, Entry, Exit, 1, 2, 3

TL: Initial → Entry : Call
Exit → Initial
1 → Exit : Synch 1,2
2 → Exit : Synch 1,2 Ack
3 → 1
3 → 2
Entry → 3.

These lists define the state machine controller for "Fetch." The equivalent state diagram is portrayed in figure 99.

Figure 99. Controller State Diagram for "Fetch"

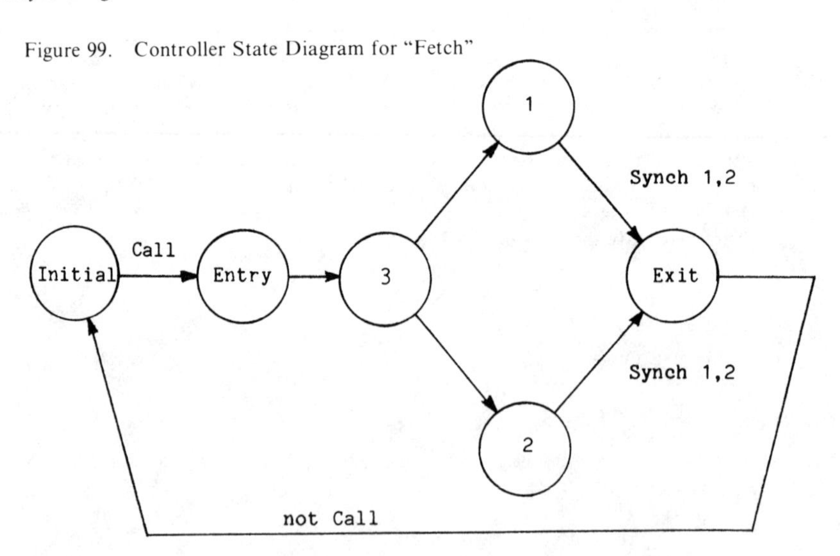

Since conditional execution and iteration have been relegated to control operators, it is possible that a control operator may require more than one controller state. By constraining the control operators to a single entry, single exit execution flow, the substates for the operator can be easily incorporated into the table. The transitions leading to and from the substates must be adjusted, of course, but this is a minor inconvenience. Incorporation of multistate control operators can be performed as a postprocessing step added to the original translation procedure.

Algorithms

Physical control is implemented using a one-hot, asynchronous state machine. Using the work of Carter and Hollaar, an SLA implementation of a given state machine can be formed from its state table. When applied to the description of a defined function, this algorithm will produce a state table for the control portion of the function implementation.

Before using algorithm GST, the algorithm RG must be called to rewrite the d-n description of the defined function (F) as a table of nodes and precedence relations (PT) (fig. 100). This table has an entry for each node in F plus the two special nodes "Entry" and "Exit" which replace the function inputs and outputs. Each table entry identifies the nodes which produce data values (precedes) for the node represented by the entry. In a final step, the nodes which require a machine state for their operation are tagged. This information is supplied by the technology database, d-tech-db.

GST produces a list of controller states (SL) and a list of state transitions (TL) by scanning the precedence table from "Exit" back to "Entry." The node list NL remembers which nodes in F require processing.

Figure 100. Fields in an Entry from Table PT

Field	Contents
Identifier	Node identifier
State-flag	Needs machine state flag
State-id	State identifier
Predecessors	Precedence list (list of producers)
Successors	Succession list (list of consumers)

Algorithm GST. (Generate the state table for a defined function.) Inputs: PT: Precedence table from RG. d-tech-db: The technology database. Outputs: SL: The list of states. TL: The list of transitions. Variables: NL: A list of nodes. INL: An intermediate list of nodes. N,IN: Node variables.

1. [Initialize.] Put Initial, Entry, and Exit in SL. Put "Initial → Entry : Call" and "Exit → Initial : not Call" into TL. Put "Exit" into NL.
2. [Terminate?] Is "Entry" the only node in NL? If yes, then terminate.
3. [Select a node.] Select and remove a node N from NL.
4. [Does it need a state?] Does this node need a machine state? If not, then ignore it and go to step 3. Otherwise, continue.

5. [Build INL for N.] Using PT, copy the precedence list for N to INL. For each IN in INL, if IN does not require state, then replace IN by its precedence list. Repeat this step until all nodes in INL require a machine state.

6. [Build states for nodes in INL.] For each IN in INL, if IN does not have an entry in SL, then put a state for it in SL and remember the identity of this state in PT.

7. [Build transitions to N.] For each IN in INL, create a transition from the state for IN to the state for N and add the new transition to TL. If the number of nodes in INL is greater than 1, add the synchronizing condition Synch to each of the new transitions with a list of all the incoming states.

8. [Add INL to NL and repeat.] Move all the nodes in INL to NL and go to step 3.

End of algorithm GST.

Algorithm RG. (Convert a defined function description to a precedence table representation.) Input: F: The d-n description of the defined function to be translated. Output: PT: The equivalent data dependency table.

1. [Initialize PT with "Entry."] Create a table entry in PT for "Entry." Assign all function inputs to this entry. The list of precedence relations for this entry is empty. The successor list points to those nodes which are connected to an input port.

2. [Initialize PT with "Exit."] Create a table entry for "Exit." Assign all function outputs to this entry. Find all nodes that produce the function outputs and remember their identities in the precedence list. The successor list is empty.

3. [Create all other PT entries.] For each node N in the graph of F, create an entry in PT and fill its precedence list with the names of the nodes that produce data values for N. Perform a similar action for the nodes that succeed N except put their identities in the successor list for N.

4. [Set state flag.] Find the d-tech-db entry for the node. If the template indicates that a machine state is needed, set the machine state flag for the entry to true, create a unique state identifier, and save it in the entry for N. Otherwise, just set the state flag to false.

End of algorithm RG.

Appendix D

Wiring

Discussion

In order to discuss system wiring, several points concerning the logical to physical translation must be reviewed:

—An encapsulation is drawn as a scaled rectangle with the same shape as it appears on the surface of the chip. (The scale factor is chosen such that the layout fits the frame on which it is drawn.) The constituent subsystems are drawn as scaled rectangles, also, representing the subregions that they occupy in the physical design.

—A defined datatype is drawn as a scaled rectangle with the shape it will have when instantiated on the chip. Its constituents are defined functions, arbiters, and storage elements. The latter two elements are drawn in accordance with their d-tech-db cell templates as instantiated by the designer (or CAD system). Arbiters and storage elements have connection points to which wires must later be attached.

—A defined function is drawn as a scaled rectangle. All its constituents (e.g., nodes, inputs, and outputs) are instantiated circuit cells from the technology database. They are each symbolized in the plan as a scaled rectangular boundary with connection points distributed around its perimeter. Space for the state machine controller must be allocated, and it is represented by a scaled rectangle. It too has connection points for the control signals it generates and the conditions that it tests.

The full geometry of a system can be drawn by expanding each subsystem down through the hierarchy beginning with the highest level encapsulation—the root of the hierarchy. The origin of the root is called the chip origin. In a fully expanded drawing of the system, the position of each defined datatype is fixed with respect to the chip origin and so are their constituents. Every connection point in every circuit, therefore, has an absolute coordinate position on the chip. This property will be exploited for inter-DD wiring.

All connections (arcs) in the logical diagrams will be translated to one or more wires in the physical design. However, some additional wires will be generated that are not explicitly specified in the logical design. Most circuit elements have power and ground connections which must be made and the controller needs to be connected to the cells that it controls.

Mechanical aids must be provided to guarantee consistency between the logical interconnections and their physical representation and to help produce the power and control connections necessary for circuit operation. One of these tools, d-router, is a wire list generator and router that will produce from the logical design and its associated floor plans a list of all explicit and implicit physical connections. Each connection is described in terms of the physical connection points at either end which must be joined by a circuit path. Using this list and the chip floor plan, d-router places the wires between the connection points, but around the islands formed by the circuit cells. As an alternative, the wire list can be given to a manual routing tool, d-wire-editor, which guides the designer through the routing process. The editor relies upon the inventiveness of the designer to find suitable channels for interconnection.

Wiring is specified at three different levels of system design and, not coincidently, they correspond to encapsulations, defined datatypes, and functions. Function wiring, at the lowest level, is the best place to start the discussion because it introduces several notions that are important for interconnection at the higher levels.

Each arc in a function graph must be translated to one or more wires. Each function input, output, and node port in a function specification is represented in the physical plan as a set of connection points. The positions of these points are defined with respect to the origin of the function plan. Through the spatial placement of the function, they are defined relative to the enclosing defined datatype plan. By locating the connection points associated with the input, output, or port at either end of an arc, an entry in the wire list can be generated showing the point to point connections to be made.

Control connections can be generated in the following way. Every control state has a connection point whose position is known relative to the origin of the controller geometry and the origin of the function floor plan. (The logic signal available at the point indicates whether that state is active or not.) Those nodes requiring a machine state to enable their operation have one or more connection points to receive this control signal. By retaining in the design database the identity of the control state associated with a particular node, the companion connection points can be located and a wire list entry produced.

The connection point information for a cell, which is essential to the production of the wire list, is described procedurally in the cell's d-tech-db entry. Figure 101 contains an example wire list routine from an inverter cell. When the cell is instantiated, the routine will return a nested list structure containing all of the connection points within the instance.

Each level of the structure consists of an identifier and another list. At the first level, the identifiers indicate the kind of signals to be described in the subsequent list. In the example, the first level has three sublists:

1. Inputs, the logical inputs to the cell.
2. Outputs, the logical outputs from the cell.
3. Power, ground, and Vdd connections.

The next level consists of several connection point sublists. Each sublist has an identifier, a layer name, and a list of coordinate pairs. The identifier is a mnemonic for the connection point list associated with it and must have the same name as one of the inputs (outputs or control ports) declared in the template header. This identifier defines an association from one of the logical ports to the list of physical connection points. Since there is only one logical input and output in this example, there is only one connection list under "Inputs" (labeled "Input") and one list under "Outputs" (called "Output"). The connection lists are "Width" coordinate pairs long. The notation:

```
<for i=0:Width–1 <0, i * Width + 13>>
```

is shorthand for "create a list with 'Width' coordinate pairs."

Some connections cannot be resolved within the function graph alone because they require access to elements outside the scope of the defined function. These connections originate at the

Figure 101. Wire List Routine from an Inverter Cell

```
Template     Not
Object type  Primitive function
Parameters   Command, Name, Width.

Inputs:      Fixed. Input.
Outputs:     Fixed. Output.
Control:     None.

Routine Connection

    if Gain <= 4 then Height := 56
    else Height := 2 * Gain + 48 fi

    Return
        <
        Inputs,
            <
            <Input, Poly, 4,
                <for i=0:Width-1 <0, i * Width + 15>>>
            >
        Outputs,
            <
            <Output, Poly, 4,
                <for i=0:Width-1 <36, i * Width + 27>>>
            >
        Power,
            <
            <Vdd,    Metal, 8,
                <<28, 0> <28, Height * Width - 8>>>
            <Ground, Metal, 8,
                <<4,  0> <4,  Height * Width - 8>>>
            >
        >
```

nodes which call a primitive datatype function or a defined function. The wire generation process must then move up one level to the defined datatype structural diagram which shows the interconnections between the defined function and its environment.

A defined function can be connected to other components in any of three ways. First, it may be made available at the subsystem interface for use by some other DD. This is represented symbolically in a defined datatype structural diagram by a connection arc from the function to an interface point on the diagram boundary. Next, a defined function may be connected to one or more primitive datatype objects, possibly through an arbiter, as the consequence of a primitive datatype function call in its graph. Symbolically, this is drawn as a connection from the function to either an arbiter or to an instance of the primitive datatype itself. Finally, the defined function may be connected with another defined function as the result of a call to that function from its graph. If the called function belongs to the same DD as the caller, the interconnection can be resolved immediately. (This will be shown as a connection between the two functions in the DD diagram.) When the called function is part of another defined datatype, it must be connected to the receptor function in that DD. This situation is represented by a connection between the function and the diagram boundary. In this case, wire resolution passes to the next higher level of the design hierarchy.

The adjectives "local" and "global" as applied to wires can now be defined precisely. A wire (or arc) that has its source and destination within a single defined datatype is said to be local because it does not extend beyond the logical or spatial scope of a particular defined datatype. Any arc or wire extending beyond the logical and spatial scope of a defined datatype is global. The interconnections between two defined datatypes are global since they cross the scope of the defined datatype at either end.

The generation of the remaining local wires is straightforward. The physical position of each primitive datatype instance, arbiter, or defined function is known relative to the origin of the defined datatype. This information is shown in the DD floor plan. In addition, the position of every call site is known as portrayed in the function plans. Through one level of geometric transformations, their positions relative to the DD origin are known, too. By matching the physical connection points according to the logical interconnection of the ports associated with them, the wire list (relative to the origin of the defined datatype) can be produced.

The global wire list is produced through successive expansion of the logical and physical design hierarchies in the following way.

—In the logical hierarchy, every instance of a subsystem in an encapsulation diagram is replaced by its own structural diagram.

—In the physical hierarchy, every subsystem is replaced with its own floor plan. The physical expansion proceeds one step further by elaborating the layout of the defined functions as well.

After expansion, all of the logical connections can be traced. At the defined datatype level, the interface function names (e.g., interface labels) provide the final transition from the graphical system structure to a particular defined function or its call. By locating the connection points associated with both ends of a global interconnection in the expanded physical hierarchy, the wire list entries for that interconnection can be produced. Once the physical connection points have been found, it is simply a matter of matching the input ports of the call with the inputs of the definition and matching the output ports of the call with the outputs of the definition. Of course, the bits within each connection strip must be matched, too.

Algorithms

Part of the translation from a d-n description to its physical implementation is the derivation of the global wire list. This algorithm exploits the system structure and plans to produce a global wire list.

The generation of the wire list for the global connections is a two-phase process. During the first phase, the logical and physical hierarchies are expanded in concert and a database of call and receptor site records is created (figs. 102, 103). A call record represents the site of a defined function call. A receptor record represents a particular instance of a defined function. During the expansion, an identifier is assigned to every connection arc. These identifiers pervade downwards in the expansion until an identifier can be assigned to either a receptor or a call. These final associations are retained in the receptor and call records. The physical receptor of a function call can be identified by finding the receptor record with the same identifier as the function call record.

Every receptor and call record pair represents a logical connection between the site of a function call and the site of an instantiated defined function. During the second phase of the wire generation

Figure 102. Call Record Structure

Field	Contents
Identifier	Internal connection identifier
Caller	Identifier of calling function
Type	Defined datatype name
Function	Name of calling function
Called	Identifier of called function
Plan	Absolute placement of call in chip plan

Figure 103. Receptor Record Structure

Field	Contents
Identifier	Internal connection identifier
Type	Defined datatype name
Function	Receptor function name
Plan	Absolute placement of function in chip plan

process the logical connection must be translated to one or more wires. Using the placement and type information present in each half of the connection record pair, the connection points for the call and the function entry and exit are generated. Then the connection points are paired in a right to left, input port to input, and output port to output fashion. The connection points within an input or output are paired from the most significant bit (MSB) to the least significant bit (LSB). Connections for control, power, "Call," and "Exit" are produced, also.

To generate the connection points from the site records, the logical and physical hierarchies must be expanded. A depth first expansion is performed with a stack that keeps track of the geometric transformations that map a particular subsystem or function into the chip coordinate space. Algorithm GDB produces a connection database that describes the function call to function instance connections. (The delay analysis procedure uses this database, also.)

Figure 104. External Import and Export Record Structure

Field	Contents
Identifier	Connection identifier
Function	Function name

Algorithm GDB. (Generate the connection database.) Inputs: S: A set of d-n descriptions. ROOT: The root encapsulation of the hierarchy. Outputs: CALLERS: List of function call records. RECEPTORS: List of function instances. ExternalImports: List of external functions used by ROOT. ExternalExports: List of functions made available for external use by ROOT. Global: ES: Expansion stack, ExternalImports, ExternalExports, CALLERS, and RECEPTORS.

1. [Initialize.] Set ES to the identity matrix. Assign a unique identifier to each of the connections in the structural diagram of ROOT. If a connection is attached to an interface, the identifier of the connection is assigned to the interface.

2. [Process external references.] For every external function called by ROOT, create and append an external import record to the ExternalImports list. For every interface function made available by ROOT, create and append an external export record to the ExternalExports list.

3. [Expand subsystems.] Using the algorithm ESS, expand each of the subsystems in ROOT. With each call to ESS, pass the type name of the subsystem, its placement transformation, and the connection identifiers that are associated with each of its interfaces.

End of Algorithm GDB.

Algorithm ESS. (Expand subsystem.) Inputs: S: A set of d-n descriptions. TYPE: Name of the subsystem type to be expanded. PLAN: Placement transformation. IDENTIFIERS: List of connection identifiers (one for every interface point.) Outputs: CALLER and RECEPTOR records. Global: ES, CALLERS, RECEPTORS.

1. [Push transform.] Concatenate (matrix multiply) PLAN with the top of ES and push the new placement transformation on ES.

2. [Test for DD or encapsulation.] If the subsystem is a defined datatype, then go to step 4.

3. [Encapsulation step.] Assign a connection identifier to each connection in the structural diagram in the following way. If a connection is connected to an interface point, it assumes the identifier of the interface. Otherwise, a new unique identifier is created and assigned to the connection arc. Apply ESS to each of the subsystems within the encapsulation using its type, placement, and identifier to interface associations. Go to step 6.

4. [Defined datatype step.] Assign a connection identifier to each connection arc in the structural diagram in the following way. If an arc is connected to an external interface point, it assumes the identifier of the interface. Otherwise, a new unique identifier is created and assigned to the connection. For each defined function within the defined datatype, perform step 5 using the function graph, placement, and identifier to interface associations of the function. When finished, go to step 6.

5. [Defined function step.] Concatenate and push the placement transformation for the defined function. Create and append a RECEPTOR record for the defined function to the RECEPTORS list. For each function call in the function graph, create and append a CALLER record to the CALLERS list. Return to the processing loop in step 4.

6. [Pop.] Pop ES, discarding the top transformation and return.

End of algorithm ESS.

Algorithm GGWL first invokes GDB to produce the connection database for a set of d-n descriptions. It then calls GW to generate the global wire list from the connection database.

Algorithm GGWL. (Generate a global wire list.) Inputs: S: A set of d-n descriptions. ROOT: The root encapsulation of the hierarchy. Output: GWL: The global wire list.

1. [Phase one.] Call algorithm GDB with S and ROOT.

2. [Phase two.] Call algorithm GW with the connection database, consisting of CALLERS and RECEPTORS.

End of algorithm GGWL.

Algorithm GW scans through the list of defined function calls, CALLERS, and matches each call with a receptor (a defined function instance). It finds the signal sources and sinks at the call and the receptor. GW then calls PCP to produce the wire list pairings.

Algorithm GW. (Generate wires.) Input: CALLERS: List of function calls. RECEPTORS: List of function instances. Output: Global wire list. Variables: C,R.

1. [Scan through CALLERS.] If CALLERS is empty, then stop. Otherwise, remove the first CALLER record, C, from CALLERS.
2. [Find RECEPTOR.] Scan RECEPTORS and find R, the RECEPTOR record with the same connection identifier as C.
3. [Find call site connections.] Push the call site transformation for C. Trace the arcs leading to the call (input ports) back to their signal sources. Trace the arcs leading from the call (output ports) to their sinks. Locate the controller state and acknowledgment connection points. Save these connection lists in CSS. Pop the call site transformation.
4. [Find receptor site connections.] Push the receptor site transformation for R. Find the connection points for each function input and output. Find the connection points for the controller Call and Exit state signals. Save these connection lists in RSS. Pop the receptor site transformation.
5. [Pair connections.] Invoke algorithm PCP on the call and receptor site connections. Go back to step 1.

End of algorithm GW.

Algorithm PCP accepts two lists, CSS and RSS, which consist of the signal sources and sinks at the call site and the receptor site. The elements of CSS and RSS are connection lists. Logical inputs are presented first, logical outputs second, and, finally, the control signals. Inputs and outputs are arranged in their left to right order. The connection points within the lists are arranged from the MSB to the LSB.

Algorithm PCP. (Pair connection points.) Input: CSS: List of call site sources and sinks. RSS: RECEPTOR site sources and sinks. Output: CPP: List of connection point pairs.

1. [Error check.] Are CSS and RSS the same length? If they are, then go to step 2. Otherwise, a mismatch between sources and sinks will result and the erroneous condition should be reported.
2. [Terminate?] Are both CSS and RSS empty? If they are empty then return.
3. [Select and check.] Select the first connection list from CSS and the first connection list from RSS. If the lengths of these lists are unequal, a mismatch at the bit level will result. This condition should be reported.
4. [Match bits.] Remove the first connection point from each connection list. Record this point to point connection in the wire list. Repeat this step until the connection lists are empty. Then return to step 2.

End of algorithm PCP.

Appendix E

Delay Analysis

This appendix describes the delay analysis algorithms which were applied to the cache memory design.

Algorithm MCD calculates the maximum combinatorial delay between two state-bearing nodes in a function graph. The maximum delay determines the controller clock period since the signals passing through a combinatorial network must stabilize before any further operations (as enabled by the next controller state) can be performed. The clock period is twice the maximum delay allowing for the use of a two-phase, nonoverlapping clock.

MCD is applied to the network form of a defined function as produced by the control algorithm RG from appendix C.

Algorithm MCD. (Maximum combinatorial delay.) Inputs: NETWORK: Network form of the defined functions. Outputs: Maximum combinatorial delay. Variables: NODE: Node to be analyzed.

1. [Terminate?] Stop if all the nodes in NETWORK have been examined and return MAXIMUM.
2. [Scan nodes.] Remove an unexamined node entry from NETWORK and store it in NODE.
3. [State bearing?] Does NODE have a controller state associated with it? If it does not, it is a combinatorial function and has been analyzed by DOS; return to step 1.
4. [Analyze state-bearing node.] Pass NODE and NETWORK to DOS which finds the maximum delay through the successors of NODE. If the time returned by DOS plus the simple delay of NODE is greater than MAXIMUM, then this quantity becomes the new MAXIMUM. Go to step 1. (The simple delay is calculated by instantiating the d-tech-db template for NODE.)

End of Algorithm MCD.

Given a function network and a starting node, algorithm DOS finds the combinatorial delay through the nodes that succeed the starting node. Analysis terminates when DOS reaches a state-bearing successor node.

Algorithm DOS. (Delay of successor.) Inputs: START: Starting node. NETWORK: Network form of a defined function. Outputs: MAXIMUM: Maximum delay from START to any state bearing successor. Variables: NODE: A successor to be analyzed.

1. [Initial step.] Set MAXIMUM to 0. If START is state bearing, then return MAXIMUM (0) and terminate this branch of the depth first analysis procedure.
2. [Terminate?] If all the successors of START have been examined, then terminate and return MAXIMUM plus the simple delay of START. The simple delay is determined by instantiating the d-tech-db template for START.
3. [Scan successors.] Select an unexamined node from the successors of START and save it in NODE.
4. [Branch forward and check maximum.] Apply DOS to NODE. If the value returned by DOS is greater than MAXIMUM, it becomes the new MAXIMUM.

End of algorithm DOS.

Algorithm DA calculates the execution time of a defined function. It accepts two arguments:

—FUNCTION: The name of the interface function to be analyzed.

—ROOT: The name of the encapsulation or defined datatype that is the root of the hierarchy-containing FUNCTION.

Using algorithm GDB (app. D), it first creates the connection database for ROOT. The database will be used to resolve function call to receptor connections. DA finds the connection (receptor) identifier for FUNCTION in the database, and calls algorithm DOR. The time value reported by DOR is the execution time of FUNCTION.

Algorithm DA. (Delay analyzer.) Input: S: Set of d-n descriptions. ROOT: Hierarchy root (encapsulation or defined datatype). FUNCTION: An interface function defined within ROOT. Outputs: ET: Execution time of FUNCTION. Global: Connection database, d-n descriptions.

1. [Generate connection database.] Call algorithm GDB to create the connection database for the hierarchy below ROOT.
2. [Find identifier.] Find the connection (receptor) identifier for FUNCTION in the connection database.
3. [Calculate execution time.] Apply DOR to FUNCTION (through its connection identifier). The execution time, ET, is the value returned by DOR.

End of algorithm DA.

Algorithm DOR calculates the delay of a defined function through its receptor identifier. Using the connection database, DOR determines if the function is instantiated inside the ROOT hierarchy or if it is an external function called by ROOT. If it is external, DOR invokes algorithm ED. If the function is contained somewhere within ROOT, DOR finds the receptor record associated with the connection identifier in the RECEPTORS list. This record supplies the defined datatype and function name of the receptor as well as its absolute position within the physical hierarchy of ROOT. If the function does not call any other defined functions, then it is self-contained and algorithm LSP is executed. If the function calls at least one other defined function, then algorithm LTP is performed. The result of ED, LSP, or LTP is returned from DOR.

Algorithm DOR. (Delay of receptor.) Inputs: ID: Connection (receptor) identifier. Outputs: ET: Delay of the defined function (receptor function).

1. [External import.] If there exists a record with the identifier ID within the ExternalImport list of the connection database, then perform ED and return its result.
2. [Within hierarchy.] Otherwise, the receptor function must exist within the hierarchy. The description of the function is examined. If it does not call any other defined functions, then go to step 3. If it calls at least one other defined function, execute step 4.
3. [Self-contained.] Execute algorithm LSP and return its result.
4. [Dependent.] Perform LTP and return the result of its execution.

End of algorithm DOR.

Algorithm ED is executed when a receptor function is located outside the scope of the ROOT hierarchy. The user must enter an estimate for the execution time of the receptor since its specification is unavailable. This estimate is captured by the design system to keep an auditable trail of performance assumptions.

Algorithm ED. (External delay.) Inputs: EIR: ExternalImport list record. Outputs: EET: Estimated execution time.

1. [Prompt designer.] Prompt the designer for the estimated execution time of the external receptor, EET.
2. [Record and return.] Record EET in the design database and return it.

End of algorithm ED.

Algorithm LSP is invoked when a function is self-contained. Since the function does not call any other defined functions, its delay depends entirely upon its internal circuit delays. The delay is equal to the product of the clock period and the length of the longest state path through the function controller.

Algorithm LSP. (Longest state path.) Inputs: RR: Receptor record. Outputs: ET: Execution time. Variables: FUNCTION: d-n description of the receptor function. NETWORK: Network form of FUNCTION. MACHINE: Controller for FUNCTION. PERIOD: Clock period.

1. [Get the function description.] Using RR, find the d-n description for the function and save it in FUNCTION.
2. [Generate network.] Invoke algorithm RG to produce the network form of FUNCTION. Save the result in NETWORK.
3. [Find clock period.] Use algorithm MCD to find the maximum combinatorial delay within the function. Set PERIOD to twice the maximum delay.
4. [Generate state machine.] Use algorithm GST to produce the state machine controller for the function and save the result in MACHINE.
5. [Find longest state path.] Find the longest state path in MACHINE from the Entry state to the Exit state. This is the path which requires the most clock transitions to execute.

6. [Return.] Multiply PERIOD times the length of the longest state path plus one accounting for the Initial state. This is ET, the estimated execution time for the function. Return ET.

End of algorithm LSP.

Algorithm LTP. (Longest time path.) Inputs: RR: Receptor record. Outputs: ET: Execution time. Variables: FUNCTION: d-n description for the defined function to be analyzed. NETWORK: Precedence table form of FUNCTION. MACHINE: State machine for FUNCTION. PERIOD: Clock period. ASSOCIATIONS: List of call (connection) identifiers for the receptor. MAXIMUM: Maximum time delay.

1. [Get the function description.] Using RR, find the d-n description for the function and save it in FUNCTION.
2. [Generate network.] Invoke algorithm RG to produce the network form of FUNCTION. Save the result in NETWORK.
3. [Find clock period.] Use algorithm MCD to find the maximum combinatorial delay within the function. Set PERIOD to twice the maximum delay.
4. [Generate state machine.] Use algorithm GST to produce the state machine controller for the function and save the result in MACHINE.
5. [Collect calls.] Using the connection database, map each of the defined function calls made by FUNCTION to their physical receptors. Store the function name to connection identifier associations in ASSOCIATIONS.
6. [Find the longest time path.] Using NETWORK and MACHINE, examine the node associated with each machine state. If a node does not call a defined function, its contribution to the delay path is PERIOD. If a node calls one or more defined functions, instantiate the node and invoke DOR on the receptors. The result from DOR is the time delay incurred by the node. Using these node delays, find the path through MACHINE which takes the longest time to execute. The time needed to execute this path is ET, the execution time of the function.

End of algorithm LTP.

Appendix F

Cache Memory Design

This appendix is a terse presentation of the cache memory d-n description.

Figures 105 and 106 contain logical and physical descriptions of the cache memory system. The cache consists of four interconnected defined datatypes. The actual cache circuitry is concentrated in the defined datatype "Cache." The other three defined datatypes provide the external interfaces to the processing unit, the memory unit, and the shared address bus.

Figure 107 shows the expanded floor plan for "Cache-memory." The system has been expanded down to the primitive circuit components which make up the cache memory. The connection points are displayed, also.

The defined datatype "A-if" supports the external interface to the address bus (fig. 108). Each of the external interfaces in the cache memory is implemented as a primitive datatype. The storage element "Address" is an instance of the primitive datatype "ExtInput." "Address" is read by the primitive datatype function "Get." This function enables the external address information, which is available on the input pads to the chip, into an internal register.

Access to "Address" is controlled by the arbiter element "Distribute." The implementation of "Distribute" permits either function to read "Address" via "Get" and routes the value of the internal register to the caller. With the assistance of "Distribute," the two defined functions "Get-Addr-A" and "Get-Addr-B" implement shared read access to the address bus (figs. 109, 110). This example demonstrates one way to multiplex an external interface saving package pins. Without multiplexing, twice as many pins would be required to access the address bus.

Figure 111 is the expanded floor plan for "A-if." This diagram shows the placement of the primitive circuit elements within "A-if." The input interface circuitry and the address buffer registers occupy the most area. Since the input circuitry includes a buffer register to hold incoming data, the address buffer registers within the two defined functions are superfluous.

The state table for "Get-Addr-A" and "Get-Addr-B" is shown in figure 112. Since this simple state sequence appears in the implementation of several other defined functions, it has been assigned the generic name, "Trivial." (Size and quality do not vary.) A symbolic description of "Trivial" using the notation of [PetCP] is presented in figure 113. The Programmable Path Logic form of "Trivial" is contained in figure 114.

The defined datatype "M-if" implements the external interface to the memory unit (figs. 115, 116). Since the address bus is shared between the processing unit, memory unit, and cache, only the data bus from the cache to the memory unit must be implemented. Again, this interface is modeled as a primitive defined datatype. "ExtTristate" provides a tristate bidirectional bus between the memory unit and the cache. It is accessed by the primitive datatype functions "Get3" and "Put3." When "Get3" is called, the interface operates as an input. The interface operates as an output when "Put3" is called. At all other times, the interface is in the high impedance state. An internal register holds both the input and output data.

The defined functions "Get-M" and "Put-M" provide on-chip access to the memory unit interface. Their descriptions are shown in figures 117 through 120. "Get-M and "Put-M" both use the "Trivial" controller.

Figure 121 is the expanded floor plan for "M-if." As in the case of "A-if," most of the area is occupied by the tristate circuitry and the function buffer registers. The buffer registers could be eliminated since the tristate interface contains its own registers for incoming and outgoing data values.

The defined datatype "Pc-if" implements the external interface to the processing unit (figs. 122, 123). This interface is a tristate device, also.

The defined functions "Get-Pc" and "Put-Pc" provide read and write access to the processing unit interface (figs. 124-127). They show how access to a primitive datatype may be made available to another environment. "Get-Pc" and "Put-Pc" use the "Trivial" controller to sequence their execution.

Figure 128 is the expanded floor plan for "Pc-if." It suffers from the same deficiencies as "A-if" and "M-if." The function input and output registers could be eliminated since the tristate circuitry has its own buffer registers, and its signals could be distributed to the users of "Put-Pc" and "Get-Pc."

The defined datatype "Cache" actually implements the cache memory unit (figs. 129, 130). It consists of four defined functions and the content addressable memory "CAM." The expanded floor plan for "Cache" is shown in figure 131.

The defined function "Cache-read" is invoked by the processing unit when it asserts a read request (figs. 132-136). "Cache-read" attempts to find an entry for the memory address in the CAM with the primitive datatype function "CAM. Lookup." The CAM returns a bit indicating whether a match was made successfully or not. If a match was made, the value field is valid and is returned, also. Otherwise, the defined function "Memory-read" is called. The match bit performs two functions:

1. It enables the conditional execution of "Memory-read" if the address pattern was not successfully matched with a CAM entry.
2. It selects either the CAM value or the result of "Memory-read" depending on the success of the pattern match operation.

The data value selected by the primitive function "Oneof2" is returned by "Cache-read."

The defined function "Memory-read" is called when the CAM does not contain the data value requested by the processing unit (figs. 137-141). "Memory-read" first calls the external function "M-read" which causes the memory unit to read the addressed data value from its own store. "Memory-read" then calls "Get-M" to actually read the memory data from the external interface. The primitive datatype function "CAM. Push" is called to push the new data value and address pair into the CAM.

The defined function "Cache-write" is invoked by the processing unit when it asserts a write request (figs. 142-146). "Cache-write" performs two separate operations in parallel. One execution path calls "CAM.Invalidate" to look up the address in the CAM. If the look up is successful, the validity bit of the selected entry is cleared making its contents invalid. The other execution path first writes the data value to the external memory interface by calling the function "Put-M." Then, a write operation is invoked in the memory unit through a call to the external function "M-write." The values returned from "Cache-write" are discarded.

The defined function "Cache-reset" is invoked by the processing unit to invalidate all the cache entries (fig. 147). This function is called as part of a power-up initialization sequence. "Cache-reset" merely calls the primitive datatype function "CAM.Reset" and returns. The content addressable memory will clear all of its validity bits. "Cache-reset" uses the "Trivial" controller.

Figures 148 through 156 portray the cell geometry for the primitive circuits used in the cache memory design. Every cell is presented in two graphical forms: the detailed geometry and the floor plan (i.e., all but connection points suppressed) form.

Figure 105. Encapsulation "Cache-memory"

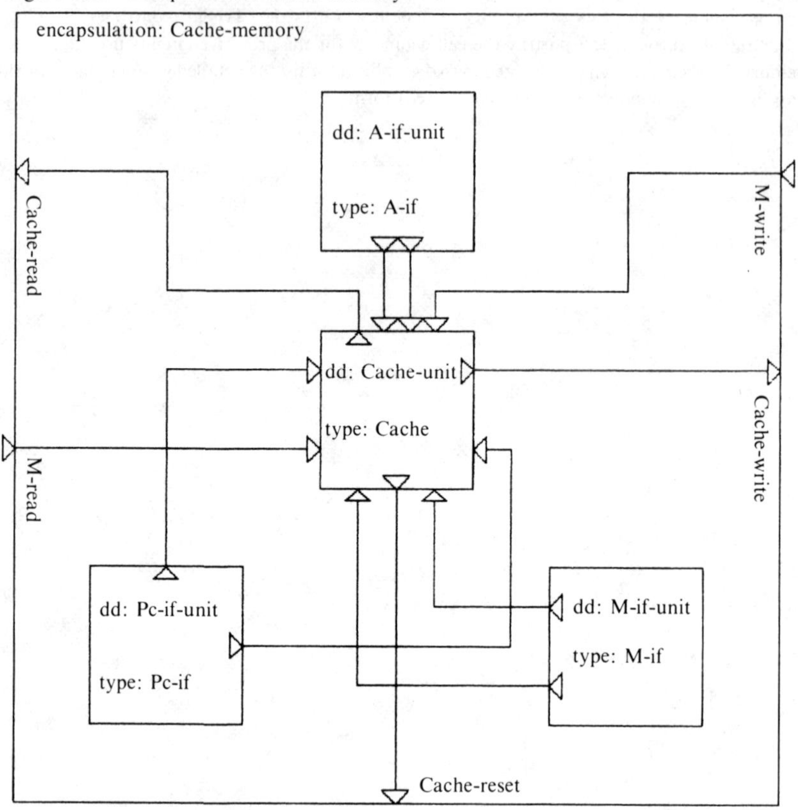

Figure 106. Floor Plan for "Cache-memory"

dd: A-if-unit type: A-if

dd: Cache-unit type: Cache

dd: Pc-if-unit type: Pc-if

dd: M-if-unit type: M-if

e: Cache-memory type: Cache-memory

Figure 107. "Cache-memory" Floor Plan with Connection Points

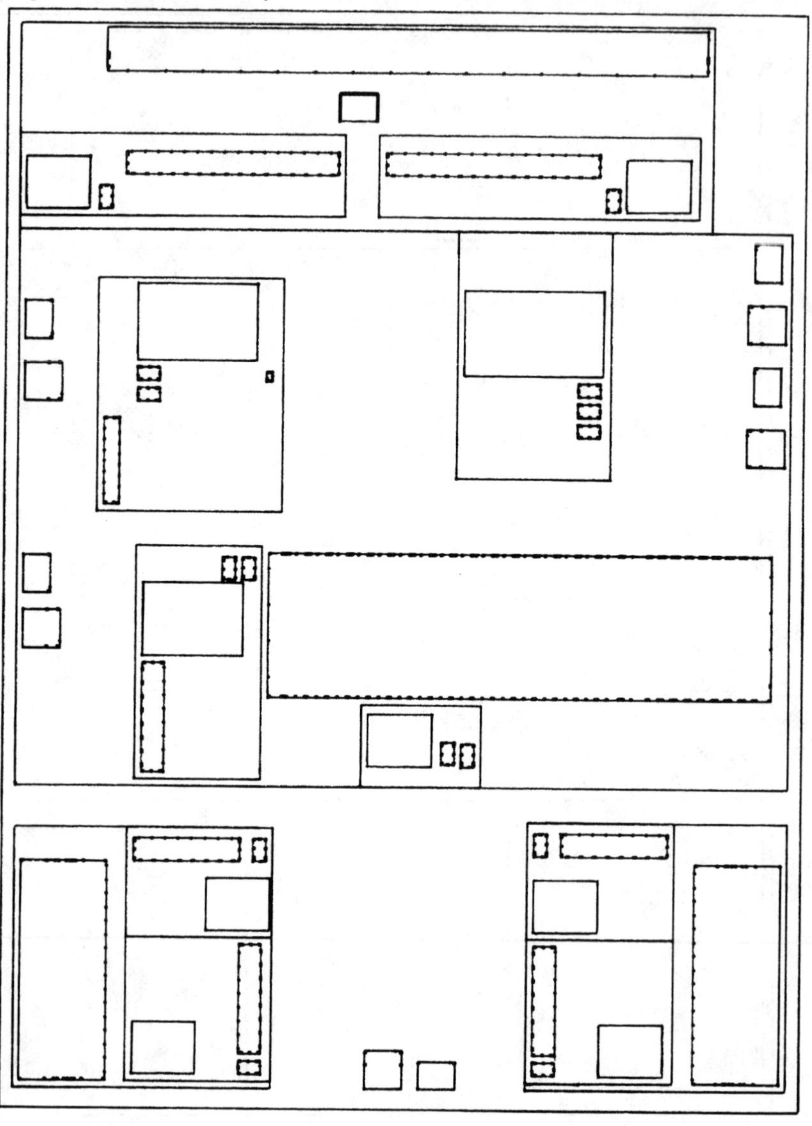

Figure 108. Defined Datatype "A-if"

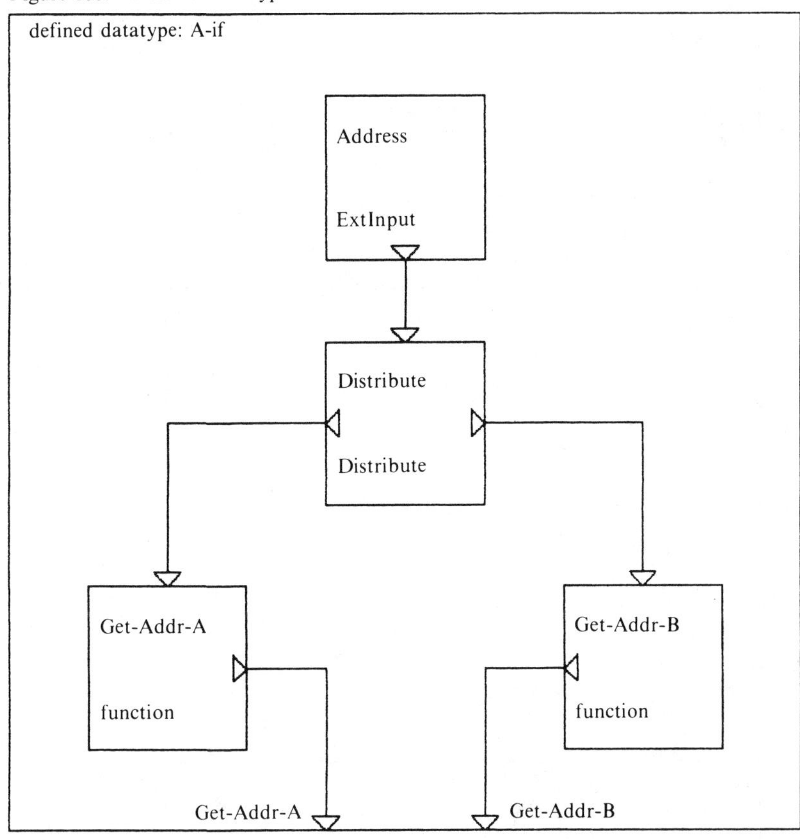

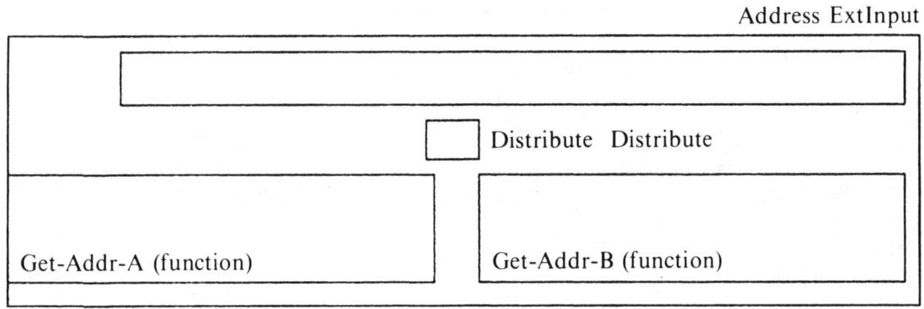

Figure 109. Defined Function "Get-Addr-A"

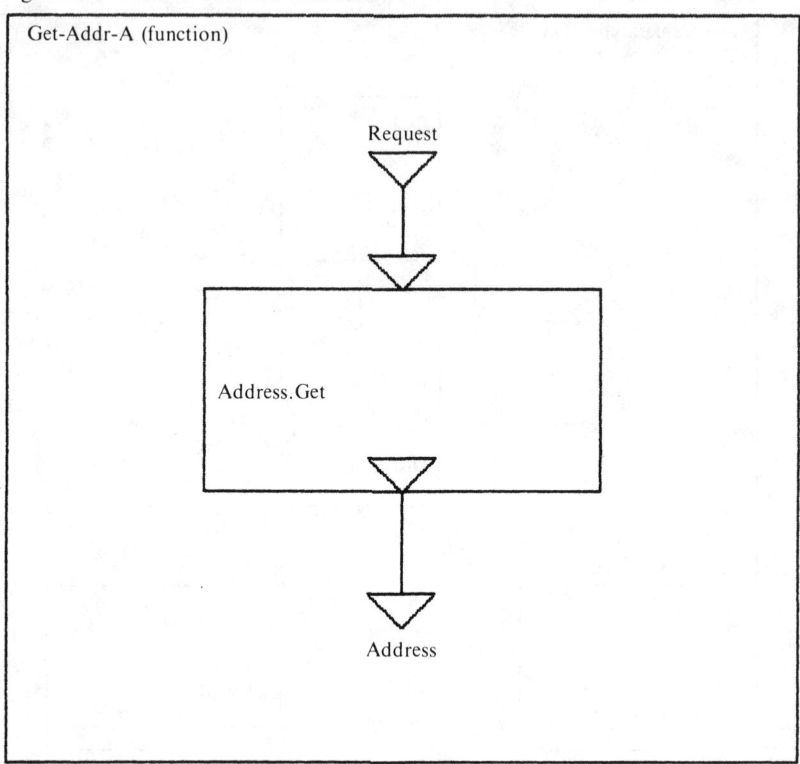

Trivial (Controller)

Address (Exit)

Request (Entry)

Get-Addr-A (function)

Figure 110. Defined Function "Get-Addr-B"

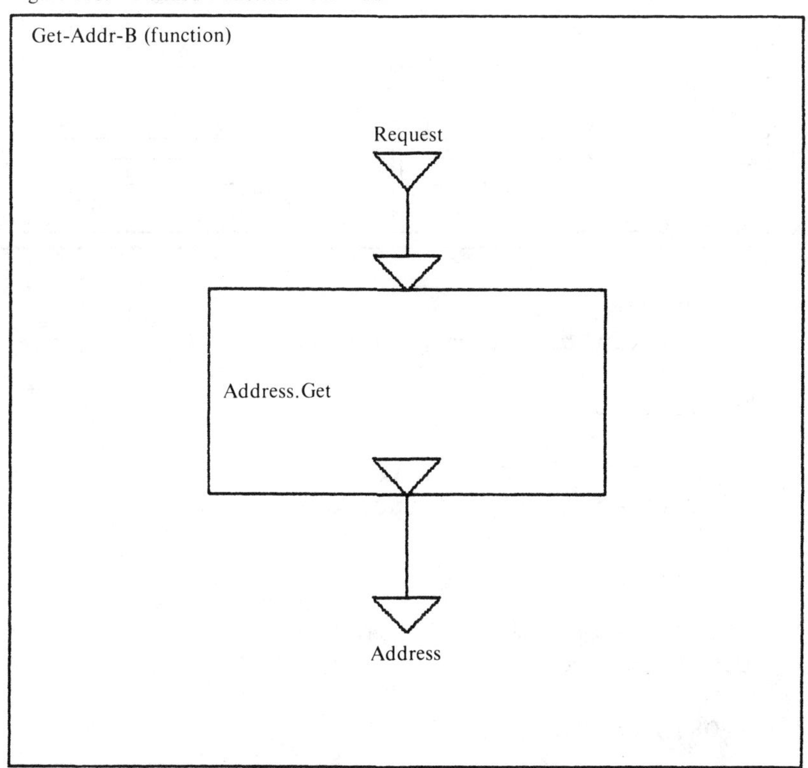

Address (Exit)

Trivial (Controller)

Request (Entry)

Get-Addr-B (function)

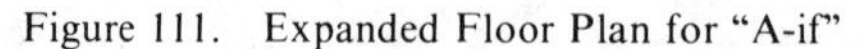

Figure 111. Expanded Floor Plan for "A-if"

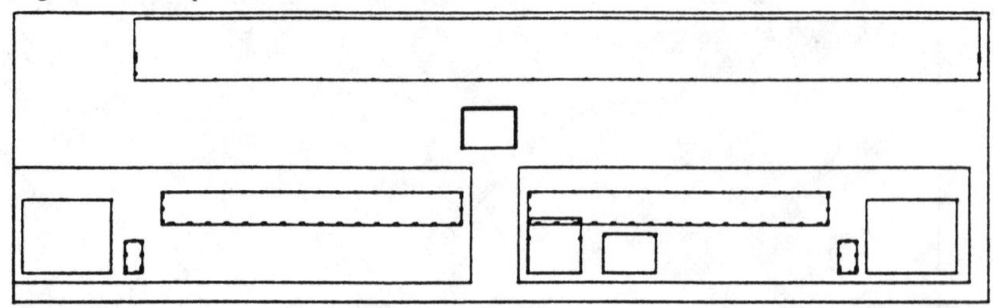

Figure 112. "Trivial" State Table

```
States: (Initial Entry Exit 1)

Transitions:

    Initial -> Entry : (Call)
    Exit -> Initial : (not Call)
    1 -> Exit :  -
    Entry -> 1 :  -
```

Figure 113. Symbolic Description of the "Trivial" Controller

```
MODULE Trivial;

IMPORTS Call,Clock;

EXPORTS Exit,S1;

VARIABLES Initial,Entry,Exit,S1;

Initial.s := Exit * Call' * Clock;
Initial.r := Call * Entry;
Entry.s := Call * Initial * Clock;
Entry.r := S1;
Exit.s := S1 * Entry * Clock;
Exit.r := Initial * Call';
```

Figure 114. PPL for the "Trivial" Controller

Controller: Trivial

Dimensions: 11 by 10

```
                                      C
                  E               C   L
                  X               A   O
                  I       S       L   C
                  T       1       L   K
----------------- * ----- * ----- * - * -
| RS FF | RS FF | RS FF | RS FF |INV|   |
|   I   |   E   |   E   |   S   | C |   |
|   N   |   N   |   X   |   1   | A |   |
|   I   |   T   |   I   |       | L |   |
|   T   |   R   |   T   |       | L |   |
|   I   |   Y   |       |       |   |   |
|   A   |       |       |       |   |   |
|   L   |       |       |       |   |   |
|=======|=======|=======|=======|===|   |
|                            ---        |
|   | S           1               0   1 |
|                ---         ---        |
|   | R   1                       1 |   |
|    ---         ---         ---        |
| 1           S                   1   1 |
|    ---         ---         ---        |
|   | . |   | R           1 | . |   |   |
|    ---     --- ---         ---        |
|   | . | 1           S   1           1 |
|    --- --- --- ---     --- ---     ---|
| 1                   R           0 | . |
-----------------------------------------
```

Figure 115. Defined Datatype "M-if"

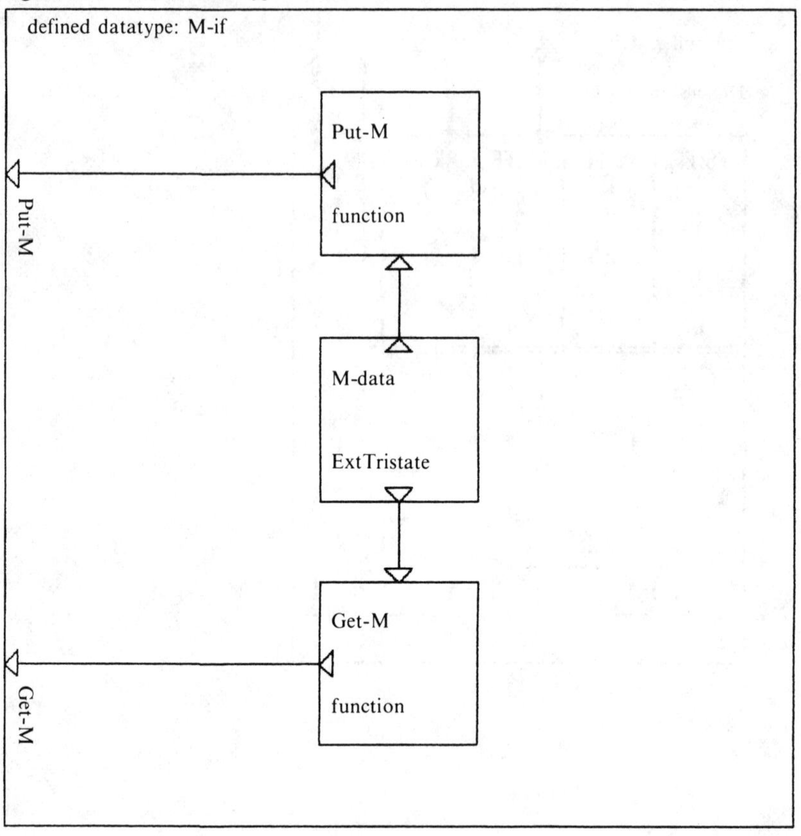

Figure 116. Plan for "M-if"

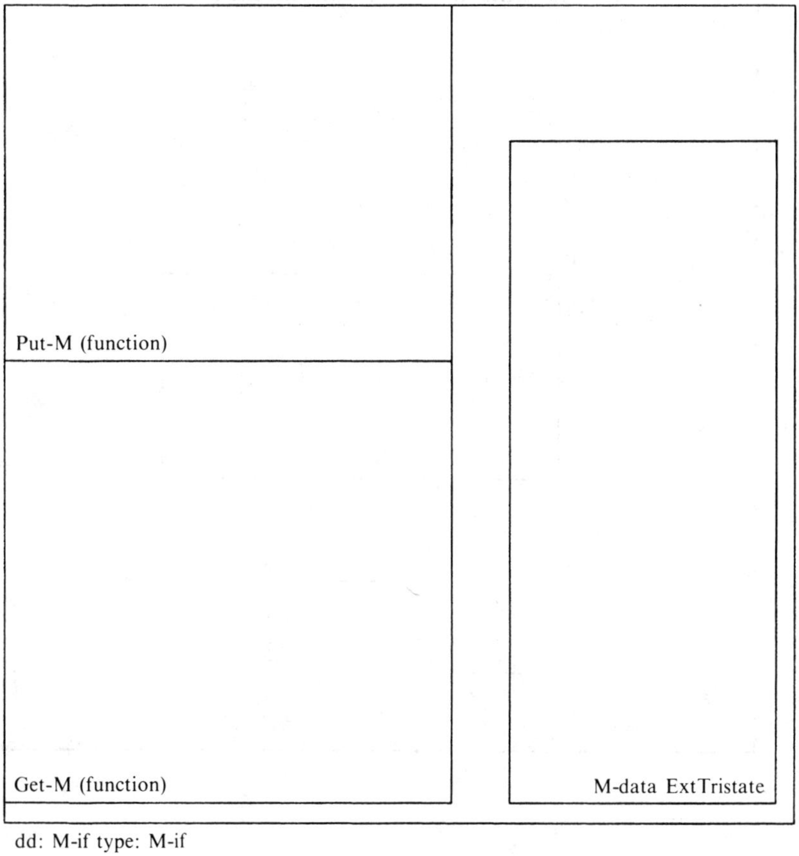

Put-M (function)

Get-M (function)

M-data ExtTristate

dd: M-if type: M-if

Figure 117. Defined Function "Get-M"

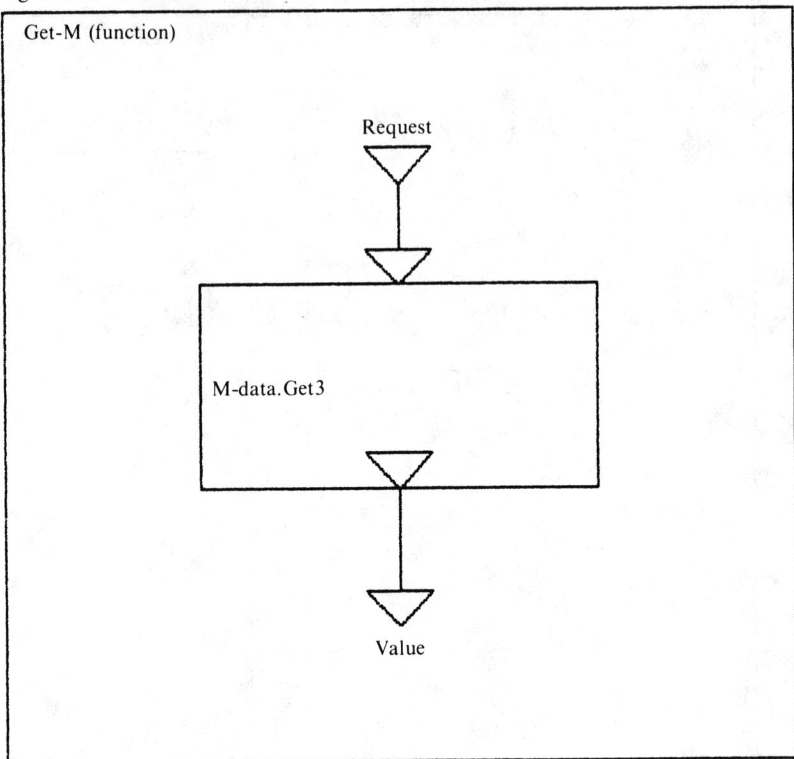

Figure 118. Plan for "Get-M"

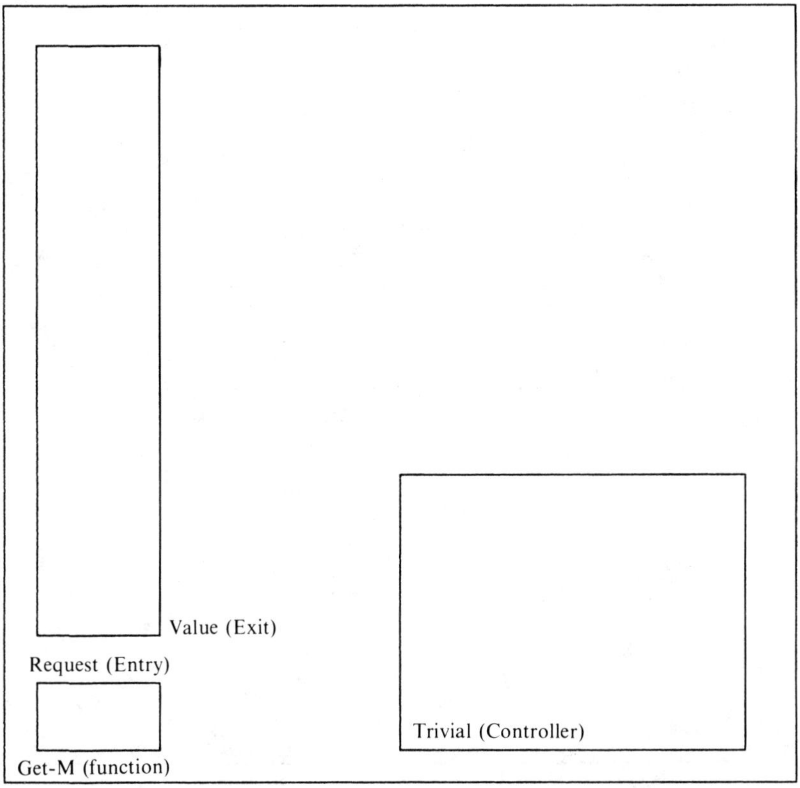

Value (Exit)
Request (Entry)
Get-M (function)
Trivial (Controller)

Figure 119. Defined Function "Put-M"

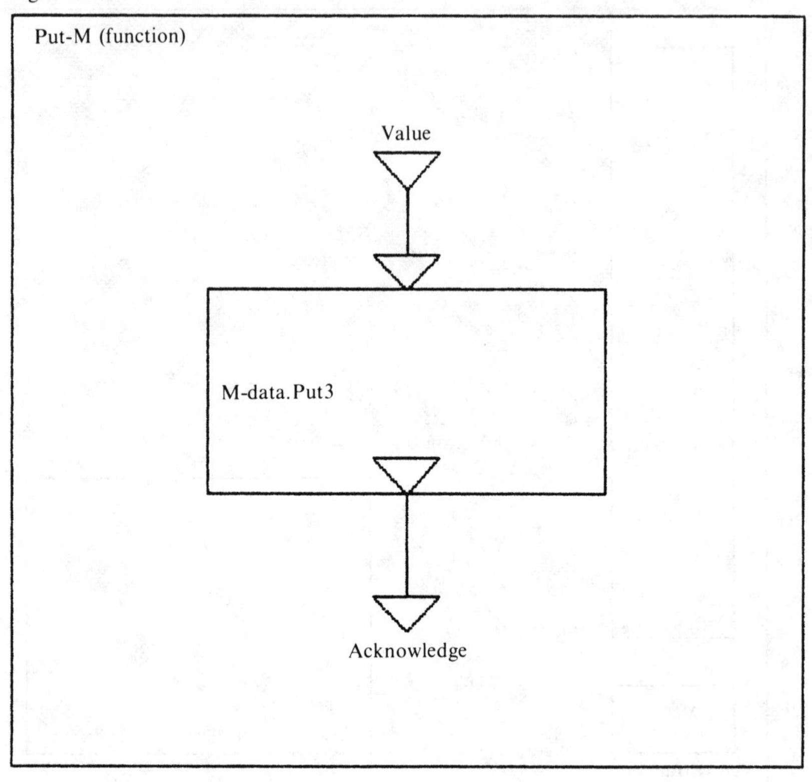

Figure 120. Plan for "Put-M"

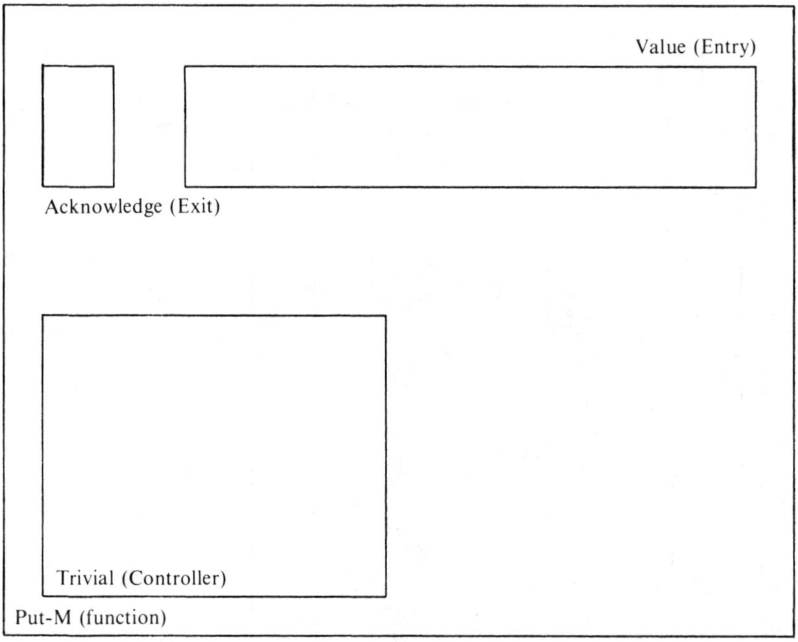

Figure 121. Expanded Floor Plan for "M-if"

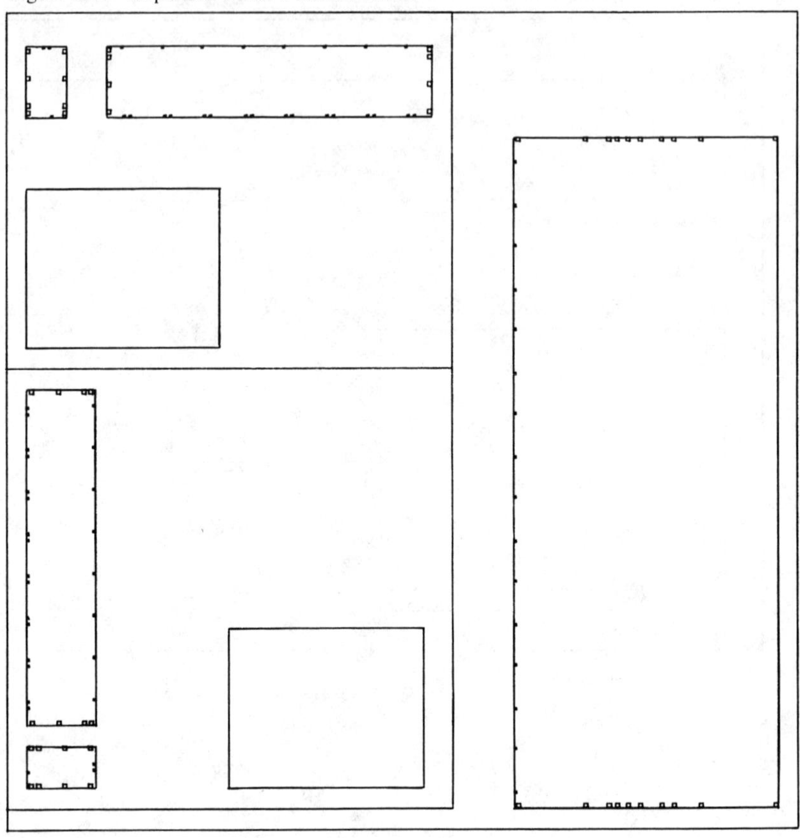

Figure 122. Defined Datatype "Pc-if"

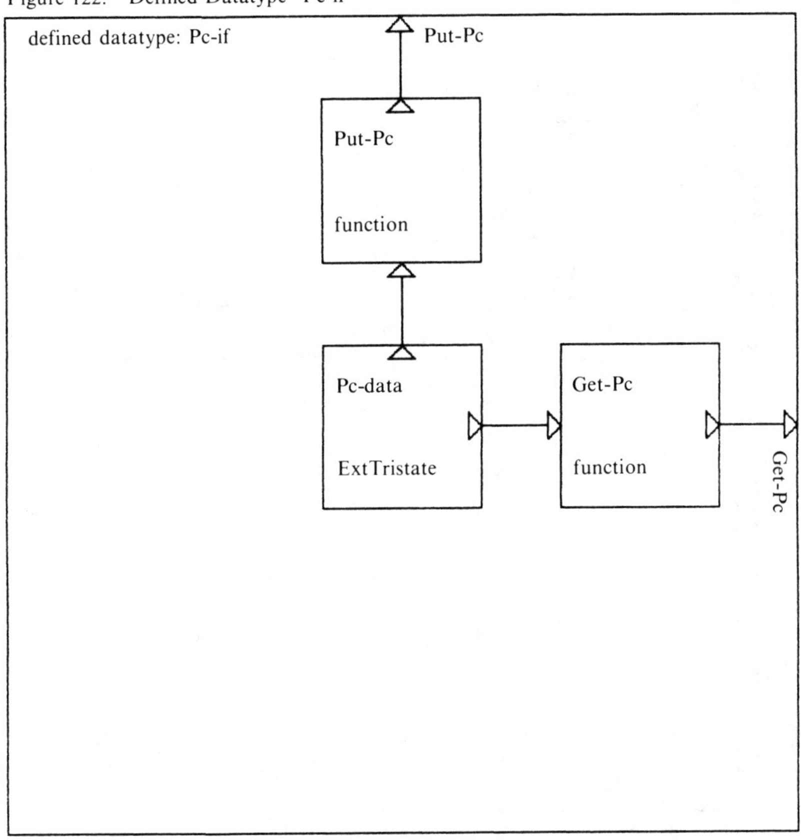

Figure 123. Plan for "Pc-if"

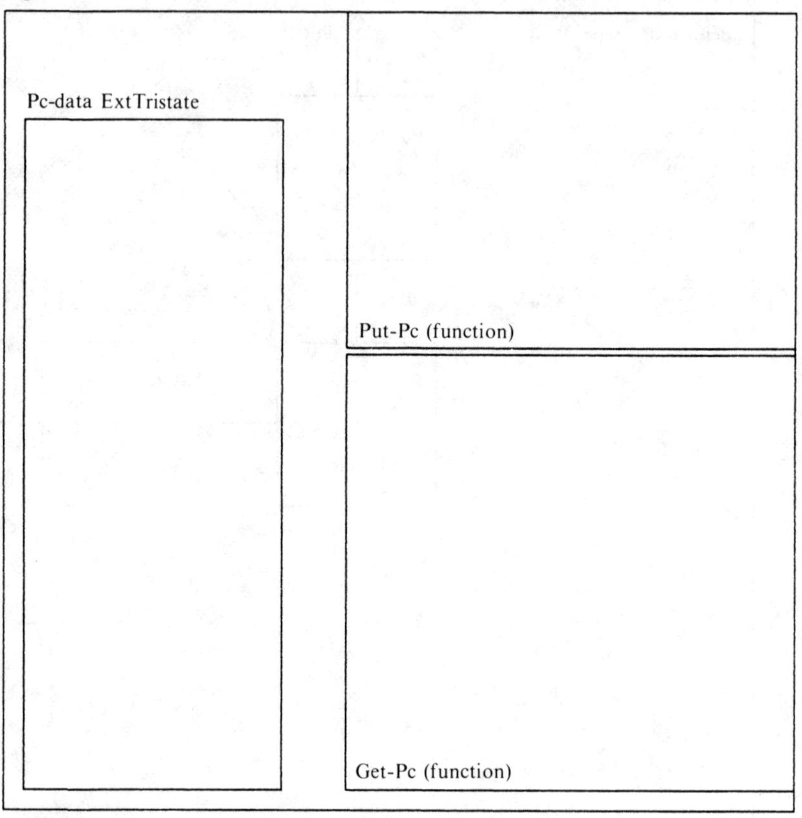

dd: Pc-if type: Pc-if

Figure 124. Defined Function "Get-Pc"

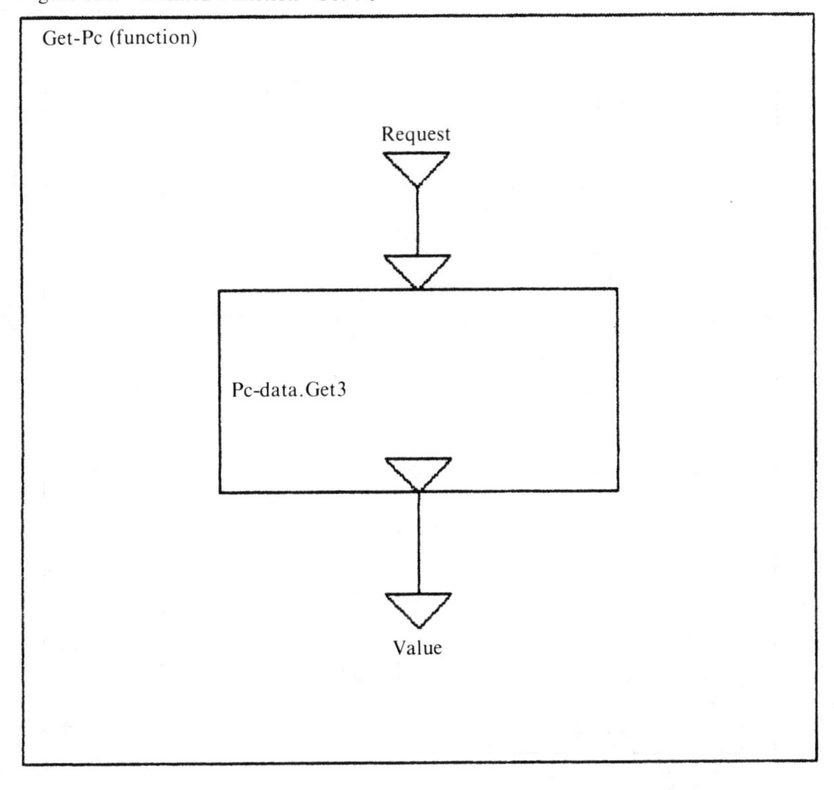

Figure 125. Plan for "Get-Pc"

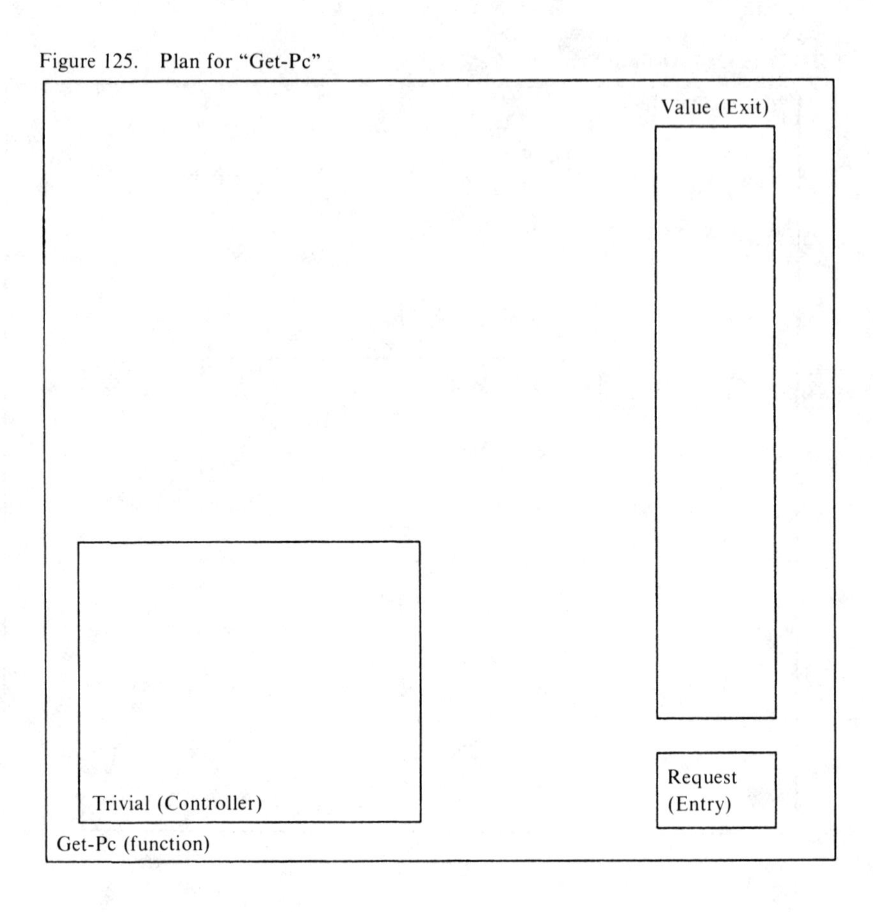

Figure 126. Defined Function "Put-Pc"

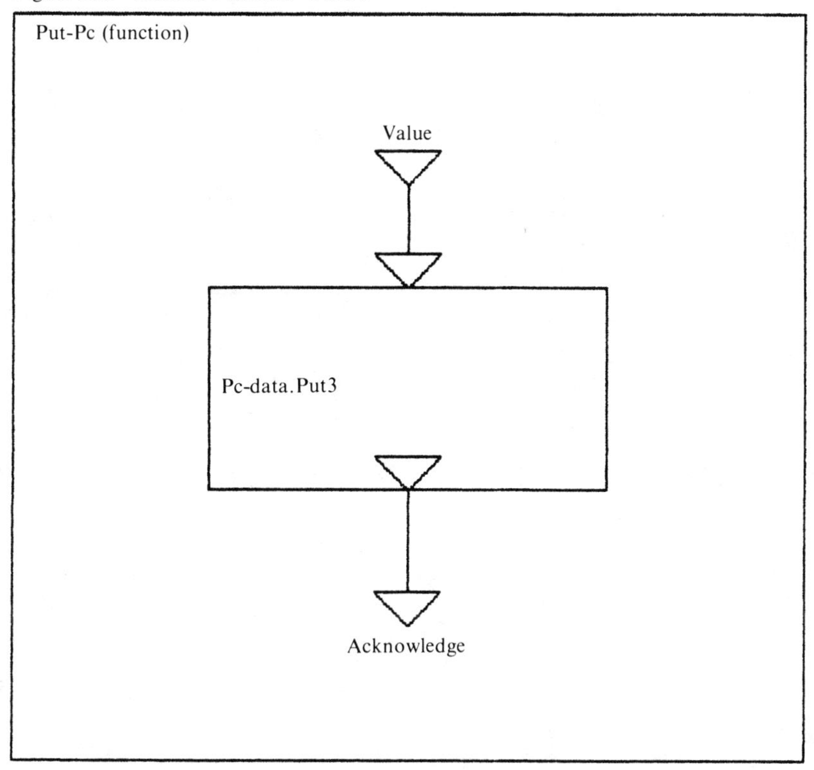

Figure 127. Plan for "Put-Pc"

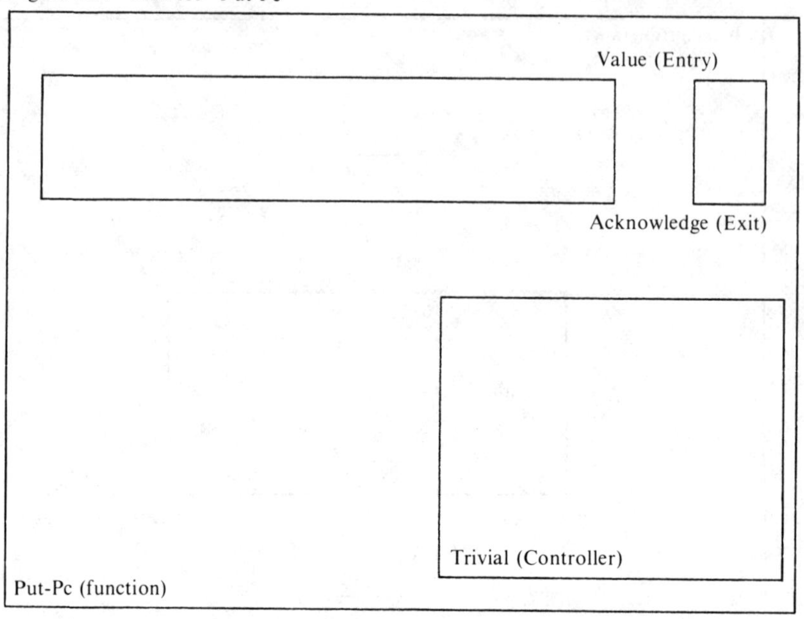

Figure 128. Expanded Floor Plan for "Pc-if"

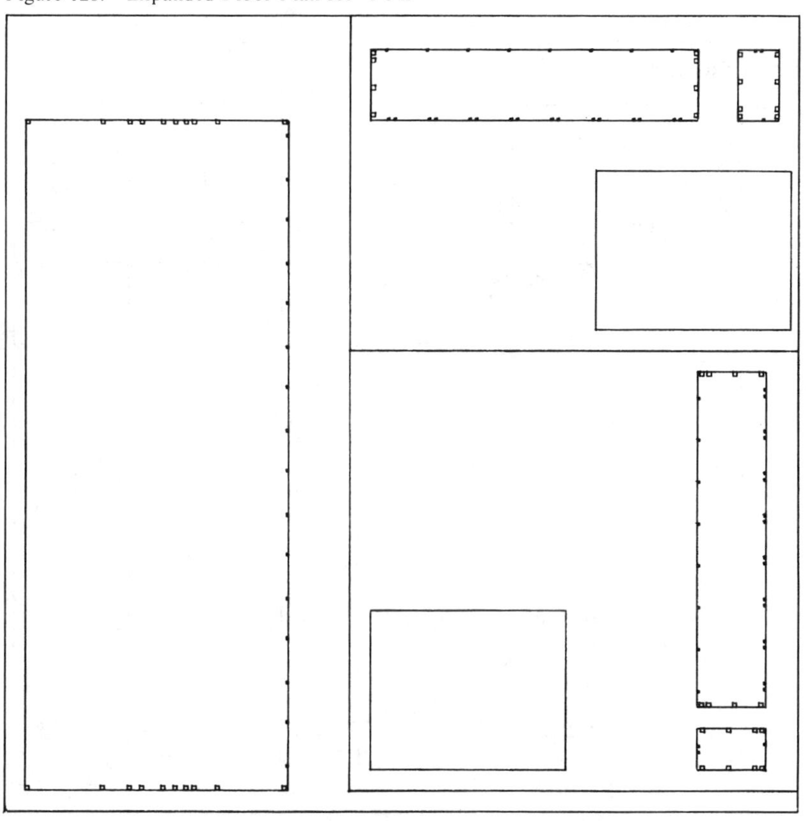

Figure 129. Defined Datatype "Cache"

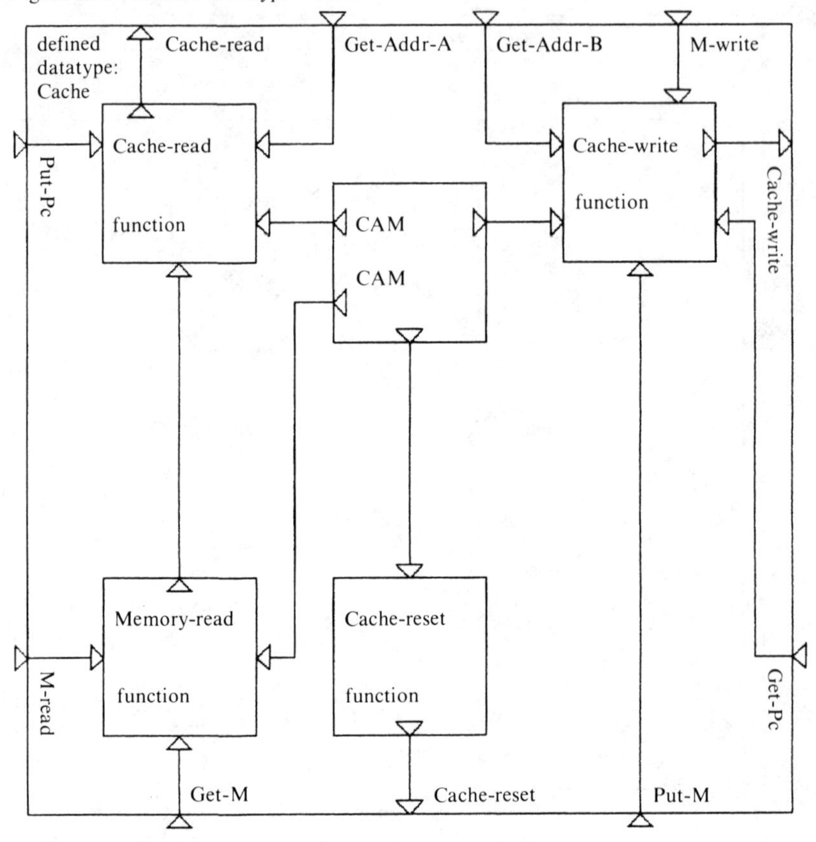

Figure 130. Plan for "Cache"

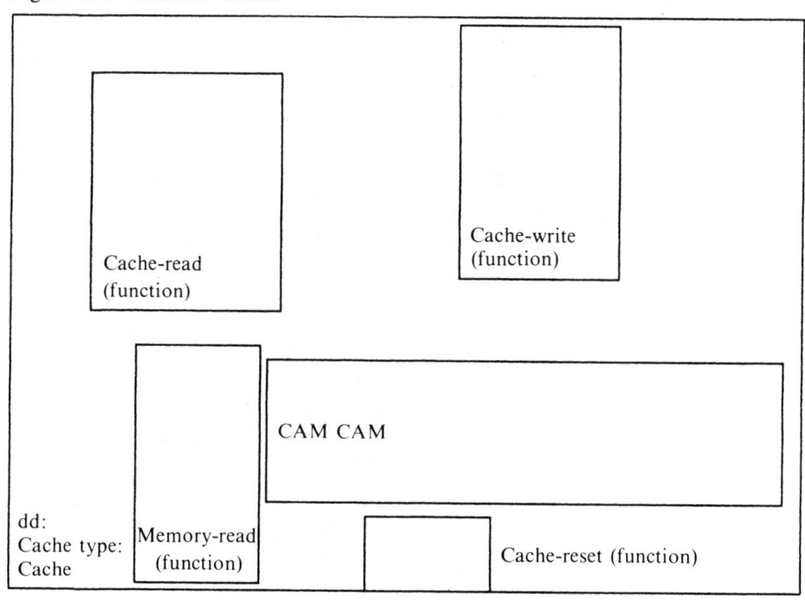

Figure 131. Expanded Floor Plan for "Cache"

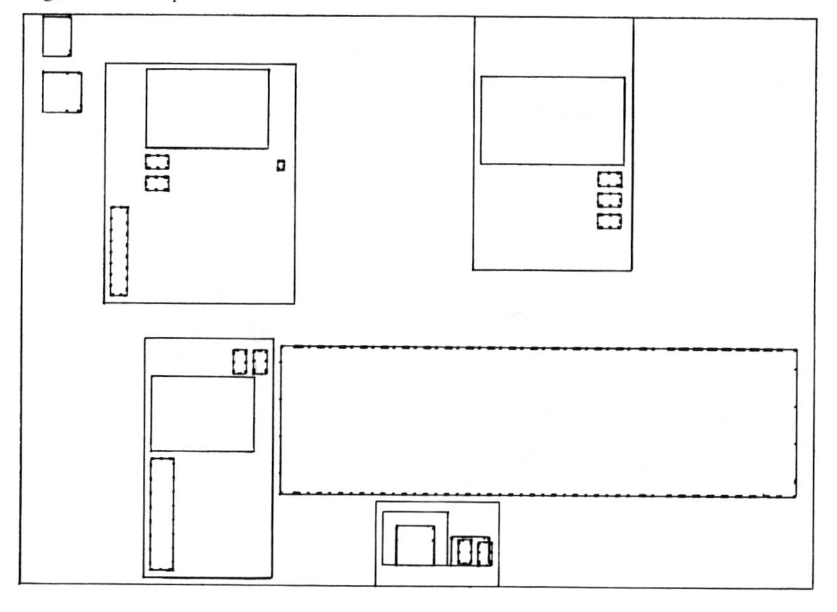

Figure 132. Defined Function "Cache-read"

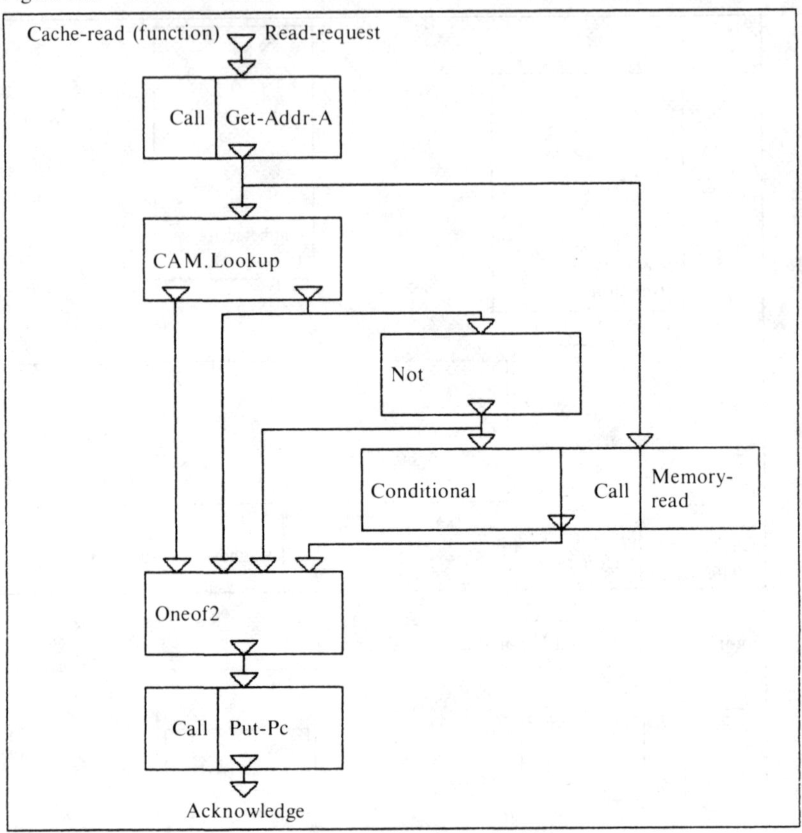

Figure 133. Plan for "Cache-read"

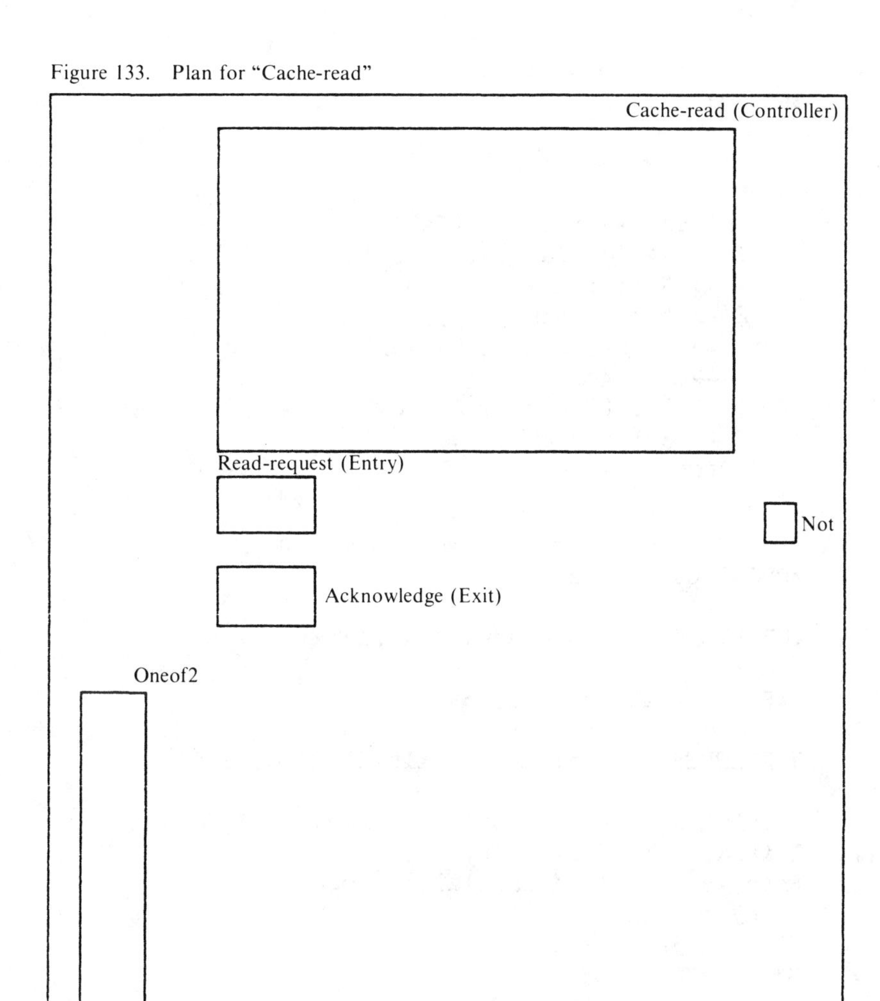

Figure 134. State Table for "Cache-read"

```
States: (Initial Entry Exit 1 2 3 4)

Transitions:

    Initial -> Entry : (Call)
    Exit -> Initial : (not Call)
    4 -> Exit : Ack
    2 -> 4 : (Synch (2 3))
    3 -> 4 : (Synch (2 3) Ack)
    1 -> 2 : Ack
    1 -> 3 : (Synch (1 2) Ack)
    2 -> 3 : (Synch (1 2))
    Entry -> 1 :  -
```

Figure 135. Symbolic Form of the "Cache-read" Controller

```
MODULE Cache-read;

IMPORTS Call,Clock,Ack1,Ack3,Ack4;

EXPORTS Exit,S1,S2,S3,S4;

VARIABLES Initial,Entry,Exit,S1,S2,S3,S4;

Initial.s := Exit * Call' * Clock;
Initial.r := Call * Entry;
Entry.s := Call * Initial * Clock;
Entry.r := S1;
S1.s := Entry * Clock;
S1.r := S2 * S3;
S2.s := S1 * Ack1 * Clock;
S2.r := S4;
S3.s := S1 * Ack1 * Clock;
S3.r := S4;
S4.s := S2 * S3 * Ack3 * Clock;
S4.r := Exit;
Exit.s := S4 * Ack4 * Clock;
Exit.r := Initial * Call';
```

Figure 136. PPL for the "Cache-read" Controller

```
                                                              C
Controller:       E                                       C   L   A   A   A
Cache-read        X                                       A   O   C   C   C
Dimensions:       I       S       S       S       S       L   C   K   K   K
17 by 19          T       1       2       3       4       L   K   1   3   4
------------------* ----- * ----- * ----- * ----- * ----- * - * - * - * - * -
| RS FF | RS FF | RS FF | RS FF | RS FF | RS FF | RS FF |INV|   |   |   |   |
|   I   |   E   |   E   |   S   |   S   |   S   |   S   | C |               |
|   N   |   N   |   X   |   1   |   2   |   3   |   4   | A |   |   |   |   |
|   I   |   T   |   I   |       |       |       |       | L |               |
|   T   |   R   |   T   |       |       |       |       | L |   |   |   |   |
|   I   |   Y   |       |       |       |       |       |   |               |
|   A   |       |       |       |       |       |       |   |   |   |   |   |
|   L   |       |       |       |       |       |       |   |               |
|=======|=======|=======|=======|=======|=======|=======|===|   |   |   |   |
|                                                                           |
|   | S           1                                       0   1 |   |   |   |
|                                                                           |
|   | R   1                                               1 |   |   |   |   |
|    ---                                                                    |
| 1           S                                           1   1 |   |   |   |
|    ---                                                                    |
|   | . |   | R           1 |   |   |   |   |   |   |   |   |   |   |   |   |
|    ---     ---                                                            |
|   | . | 1                   S                               1 |   |   |   |
|    --- --- ---                                                            |
|   | . | . | . |   |   |   | R   1       1 |   |   |   |   |   |   |   |   |
|    --- --- ---             ---                                            |
|   | . | . | . |   |   | 1           S       S               1   1 |   |   |
|    --- --- ---         --- ---                                   ---      |
|   | . | . | . |   |   | . | . |   | R       R   1 |   |   |   | . |   |   |
|    --- --- ---         --- ---     ---     ---                   ---      |
|   | . | . | . |   |   | . | . | 1       1           S       1       1 |   |
|    --- --- ---         --- --- --- --- --- ---                   --- ---  |
|   | . | . | . | 1                                   R |   |   | . | . |   |
|    --- --- --- ---     --- --- --- --- --- ---     ---         --- ---    |
|   | . | . | . | . | S                           1           1           1 |
|    --- --- --- ---     --- --- --- --- --- --- --- ---     --- --- --- ---|
| 1                   R                                   0 | . | . | . | . |
-----------------------------------------------------------------------------
```

Figure 137. Defined Function "Memory-read"

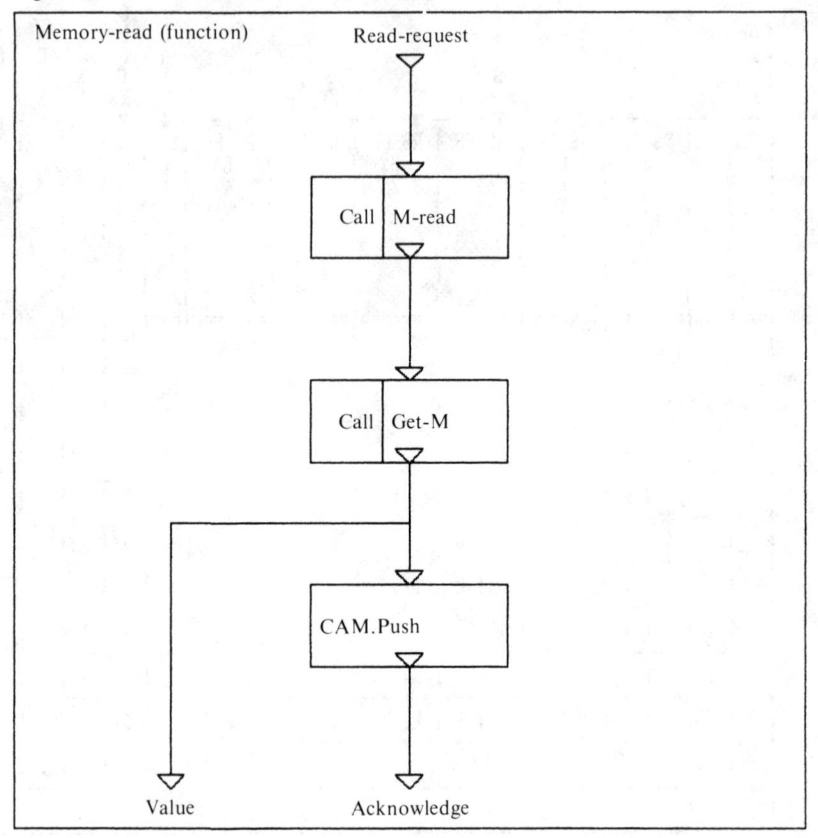

Figure 138. Plan for "Memory-read"

Figure 139. State Table for "Memory-read"

```
States: (Initial Entry Exit 1 2 3)

Transitions:

    Initial -> Entry : (Call)
    Exit -> Initial : (not Call)
    2 -> Exit : (Synch (2 3) Ack)
    3 -> Exit : (Synch (2 3))
    1 -> 2 : Ack
    2 -> 3 : (Synch (2 Entry) Ack)
    Entry -> 3 : (Synch (2 Entry))
    Entry -> 1 :  -
```

Figure 140. Symbolic Form of the "Memory-read" Controller

```
MODULE Memory-read;

IMPORTS Call,Clock,Ack1,Ack2;

EXPORTS Exit,S1,S2,S3;

VARIABLES Initial,Entry,Exit,S1,S2,S3;

Initial.s := Exit * Call' * Clock;
Initial.r := Call * Entry;
Entry.s := Call * Initial * Clock;
Entry.r := S1 * S3;
S1.s := Entry * Clock;
S1.r := S2;
S2.s := S1 * Ack1 * Clock;
S2.r := Exit;
S3.s := Entry * S2 * Ack2 * Clock;
S3.r := Exit;
Exit.s := S2 * Ack2 * S3 * Clock;
Exit.r := Initial * Call';
```

Figure 141. PPL for the "Memory-read" Controller

Controller:
Memory-read
Dimensions:
16 by 16

```
                                                      C
                  E                               C   L   A   A
                  X                               A   O   C   C
                  I       S       S       S       L   C   K   K
                  T       1       2       3       L   K   1   2
----------------- * ----- * ----- * ----- * ----- * - * - * - * -
| RS FF | RS FF | RS FF | RS FF | RS FF | RS FF |INV|   |   |   |
|   I   |   E   |   E   |   S   |   S   |   S   | C |   |   |   |
|   N   |   N   |   X   |   1   |   2   |   3   | A |   |   |   |
|   I   |   T   |   I   |       |       |       | L |   |   |   |
|   T   |   R   |   T   |       |       |       | L |   |   |   |
|   I   |   Y   |       |       |       |       |   |   |   |   |
|   A   |       |       |       |       |       |   |   |   |   |
|   L   |       |       |       |       |       |   |   |   |   |
|=======|=======|=======|=======|=======|=======|===|   |   |   |
|                                                               |
|   | S           1                               0   1 |   |   |
|                                                               |
|   | R 1                                         1 |   |   |   |
|    ---                                                        |
| 1           S                                   1   1 |   |   |
|    ---                                                        |
|   | . |   | R           1               1 |   |   |   |   |   |
|    ---     ---                                                |
|   | . | 1               S                           1 |   |   |
|    ---     ---                                                |
|   | . |   | . |   |   |   | R   1 |   |   |   |   |   |   |   |
|    ---     ---             ---                                |
|   | . |   | . |   |   | 1       S               1   1 |   |
|    ---     ---         --- ---                         ---    |
|   | . |   | . | 1             R       R |   |   | . |   |
|    ---     --- ---     --- ---     ---                 ---    |
|   | . | 1                 1           S       1       1 |
|    --- --- --- ---     --- ---     ---     ---         ---    |
|   | . | . | . | . | S           1       1           1       1 |
|    --- --- --- ---     --- --- --- --- --- ---     --- --- ---|
| 1                   R                           0 | . | . | . |
-----------------------------------------------------------------
```

Figure 142. Defined Function "Cache-write"

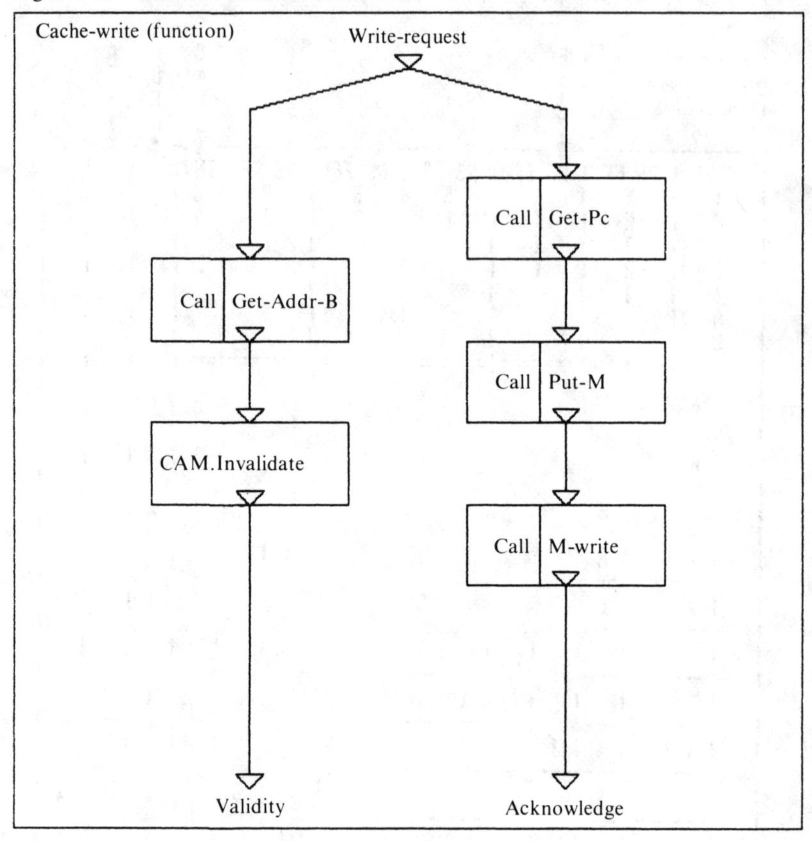

Figure 143. Plan for "Cache-write"

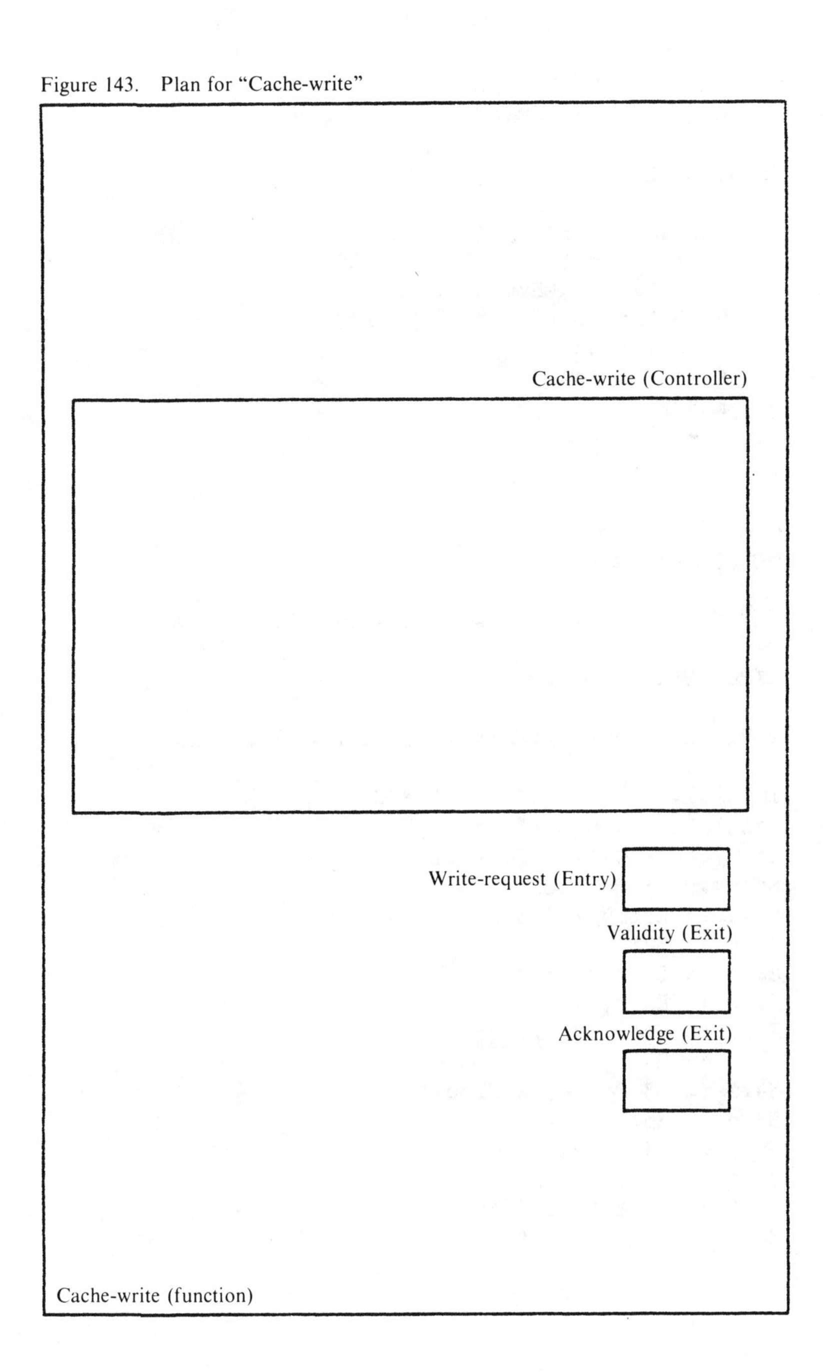

Figure 144. State Table for "Cache-write"

```
States: (Initial Entry Exit 1 2 3 4 5)

Transitions:

    Initial -> Entry : (Call)
    Exit -> Initial : (not Call)
    2 -> Exit : (Synch (2 5))
    5 -> Exit : (Synch (2 5) Ack)
    1 -> 2 : Ack
    4 -> 5 : Ack
    Entry -> 1 :  -
    3 -> 4 : Ack
    Entry -> 3 :  -
```

Figure 145. Symbolic Form of the "Cache-write" Controller

```
MODULE Cache-write;

IMPORTS Call,Clock,Ack1,Ack3,Ack4,Ack5;

EXPORTS Exit,S1,S2,S3,S4,S5;

VARIABLES Initial,Entry,Exit,S1,S2,S3,S4,S5;

Initial.s := Exit * Call' * Clock;
Initial.r := Call * Entry;
Entry.s := Call * Initial * Clock;
Entry.r := S1 * S3;
S1.s := Entry * Clock;
S1.r := S2;
S2.s := S1 * Ack1 * Clock;
S2.r := Exit;
S3.s := Entry * Clock;
S3.r := S4;
S4.s := S3 * Ack3 * Clock;
S4.r := S5;
S5.s := S4 * Ack4 * Clock;
S5.r := Exit;
Exit.s := S2 * S5 * Ack5 * Clock;
Exit.r := Initial * Call';
```

Figure 146. PPL for the "Cache-write" Controller

```
                                                                      C
Controller:       E                                               C   L   A   A   A   A
Cache-write       X                                               A   O   C   C   C   C
Dimensions:       I       S       S       S       S       S       L   C   K   K   K   K
19 by 22          T       1       2       3       4       5       L   K   1   3   4   5
------------------* ----- * ----- * ----- * ----- * ----- * ----- * - * - * - * - * - * -
| RS FF | RS FF | RS FF | RS FF | RS FF | RS FF | RS FF | RS FF |INV|   |   |   |   |   |
|   I   |   E   |   E   |   S   |   S   |   S   |   S   |   S   | C |                   |
|   N   |   N   |   X   |   1   |   2   |   3   |   4   |   5   | A |   |   |   |   |   |
|   I   |   T   |   I   |       |       |       |       |       | L |                   |
|   T   |   R   |   T   |       |       |       |       |       | L |   |   |   |   |   |
|   I   |   Y   |       |       |       |       |       |       |   |                   |
|   A   |       |       |       |       |       |       |       |   |   |   |   |   |   |
|   L   |       |       |       |       |       |       |       |   |                   |
|=======|=======|=======|=======|=======|=======|=======|=======|===|   |   |   |   |   |
|                                                                                       |
|   | S           1                                               0   1 |   |   |   |   |
|                                                                                       |
|   | R   1                                                       1 |   |   |   |   |   |
|    ---                                                                                |
| 1           S                                                   1   1 |   |   |   |   |
|    ---                                                                                |
|   | . |   | R           1               1 |   |   |   |   |   |   |   |   |   |   |   |
|    ---     ---                                                                        |
|   | . | 1                   S               S                       1 |   |   |   |   |
|    --- --- ---                                                                        |
|   | . | . | . |   |   |   | R   1 |   |   |   |   |   |   |   |   |   |   |   |   |   |
|    --- --- ---             ---                                                        |
|   | . | . | . |   |   | 1           S                               1   1 |   |   |   |
|    --- --- ---         --- ---                                         ---            |
|   | . | . | . | 1                   R                       R |   |   | . |   |   |   |
|    --- --- --- ---     --- ---     ---                                 ---            |
|   | . | . | . | . |   | . | . |   | . |   | R   1 |   |   |   |   | . |   |   |   |
|    --- --- --- ---     --- ---     ---     ---                         ---            |
|   | . | . | . | . |   | . | . |   | . | 1           S               1       1 |   |   |
|    --- --- --- ---     --- ---     --- --- ---                         --- ---        |
|   | . | . | . | . |   | . | . |   | . | . | . |   | R   1 |   |   |   | . | . |   |   |
|    --- --- --- ---     --- ---     --- --- ---     ---                 --- ---        |
|   | . | . | . | . |   | . | . |   | . | . | . | 1           S       1           1 |   |
|    --- --- --- ---     --- ---     --- --- --- --- ---     ---         --- --- ---    |
|   | . | . | . | . | S           1                       1           1               1 |
|    --- --- --- ---     --- --- --- --- --- --- --- --- --- ---     --- --- --- --- ---|
| 1                   R                                           0 | . | . | . | . | . |
-----------------------------------------------------------------------------------------
```

Figure 147. Defined Function "Cache-reset"

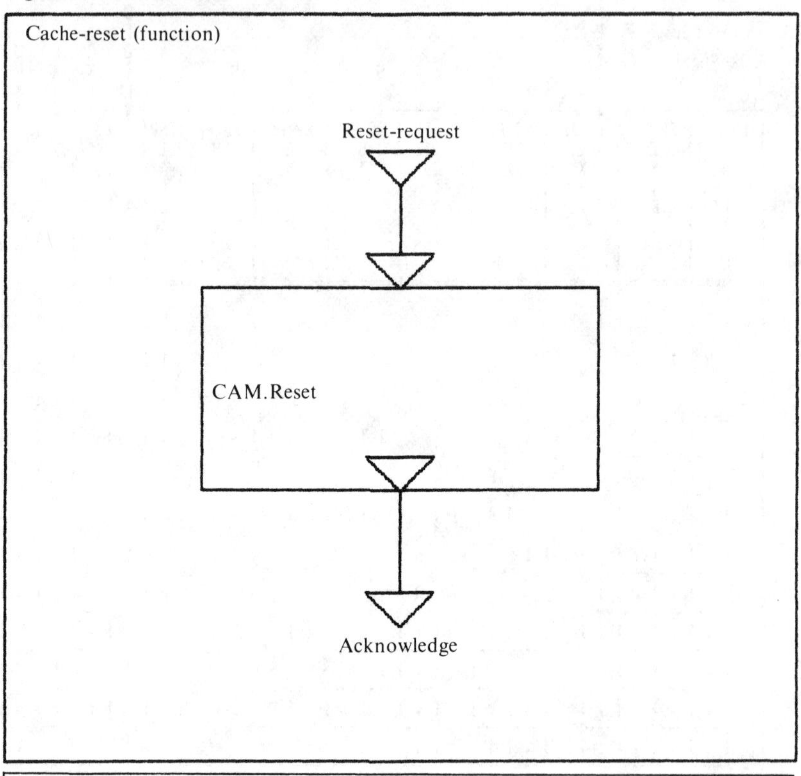

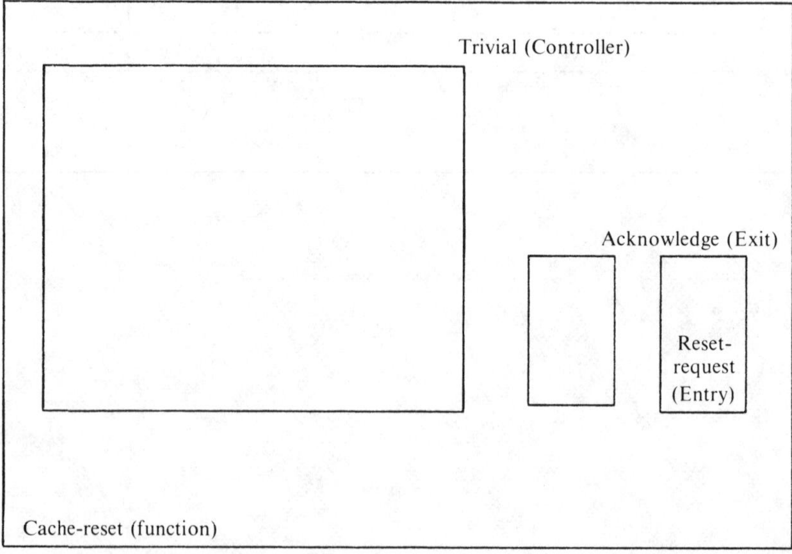

Figure 148. Content Addressable Memory

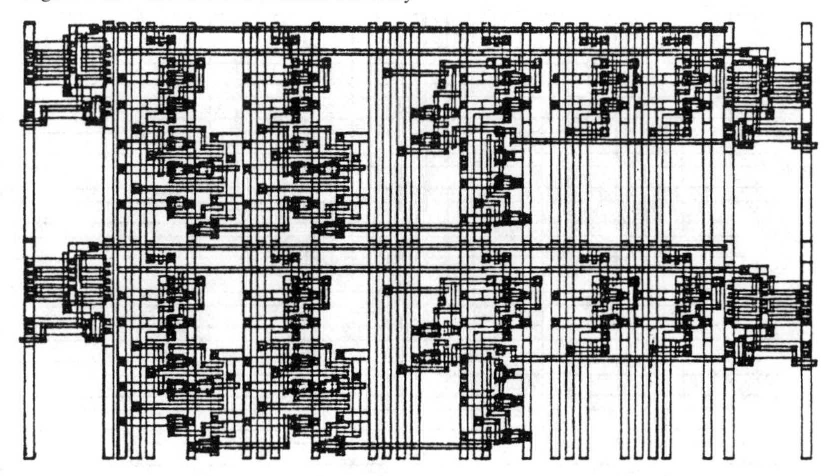

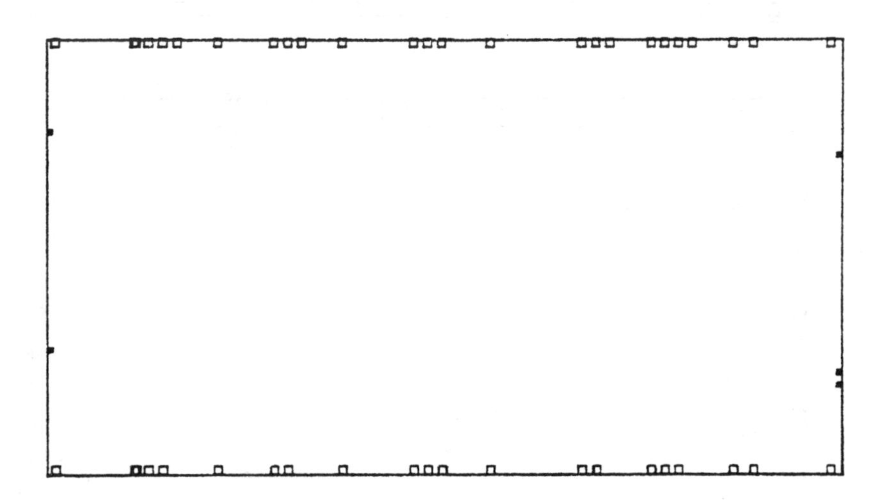

Figure 149. Distribute

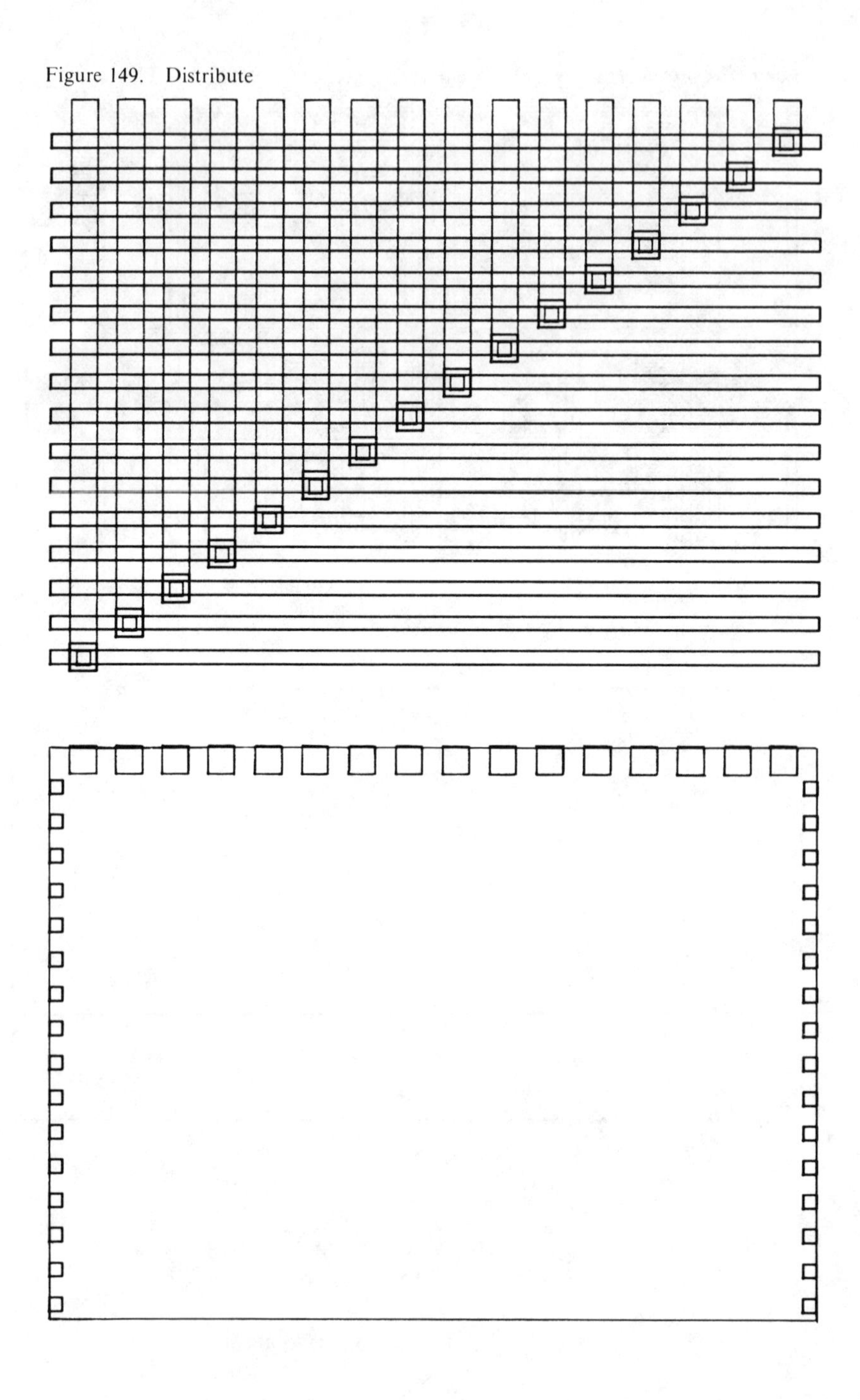

Figure 150. Latched External Input

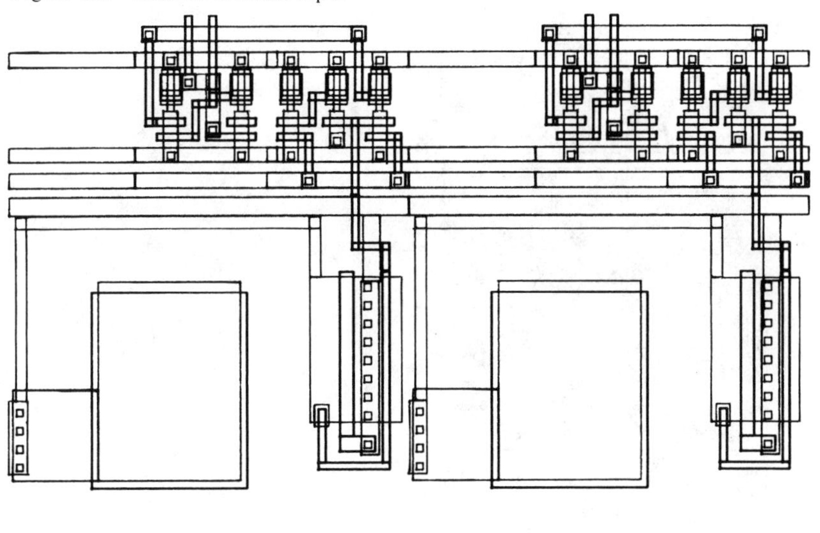

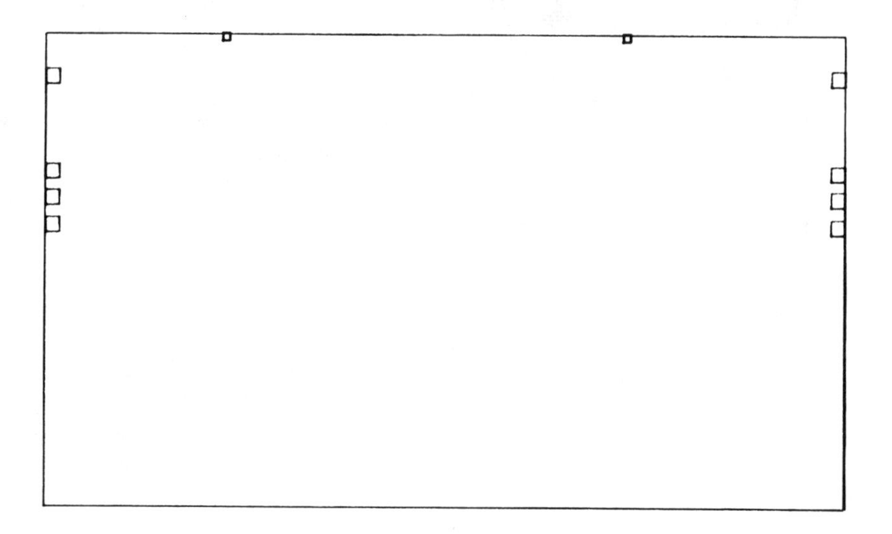

Figure 151. Latched External Tristate Interface

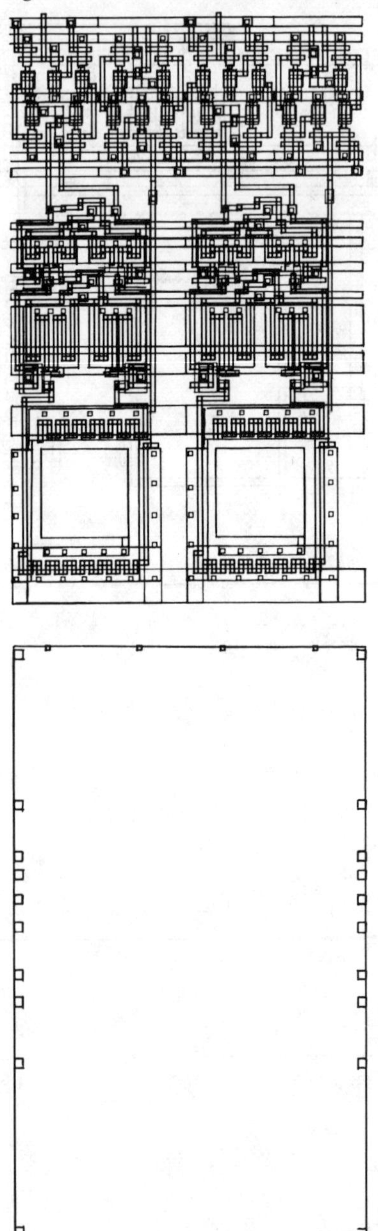

Figure 152. Signal Input

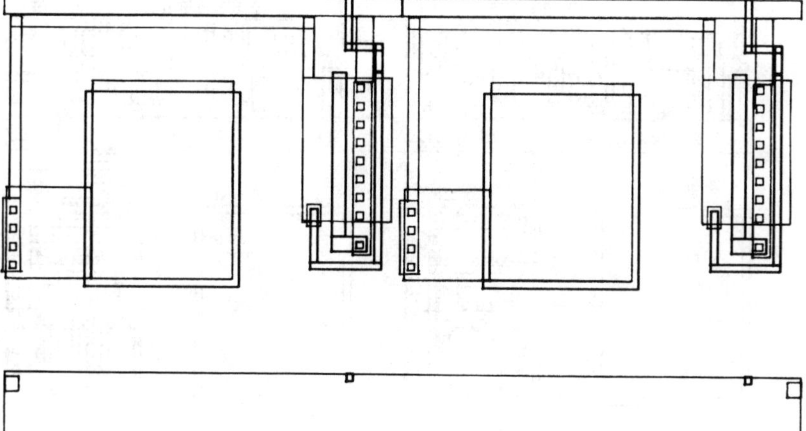

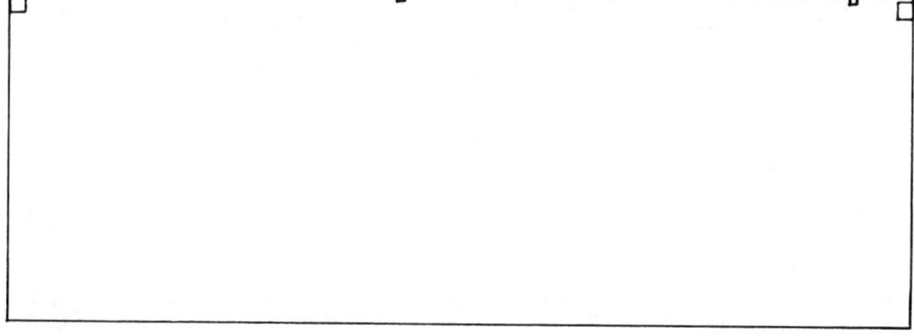

Figure 153. Signal Output

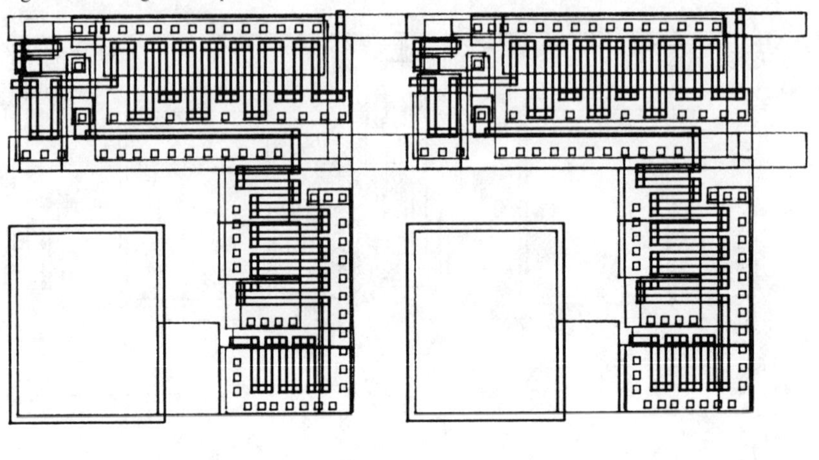

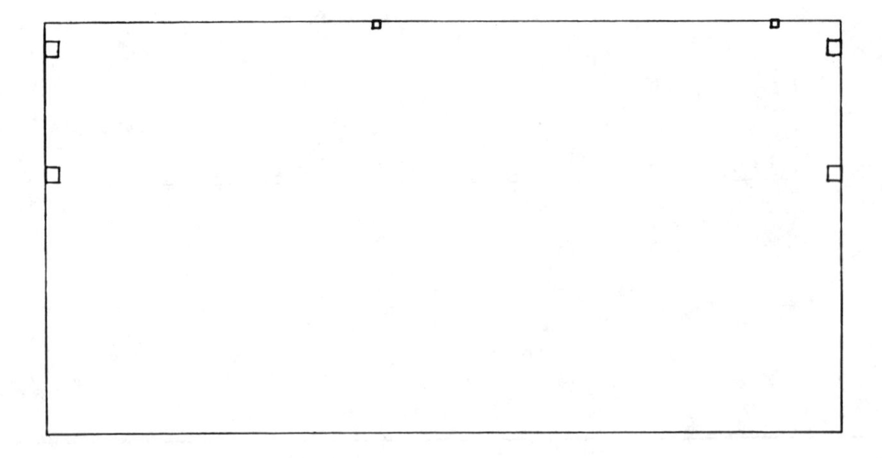

Figure 154. Entry and Exit Registers

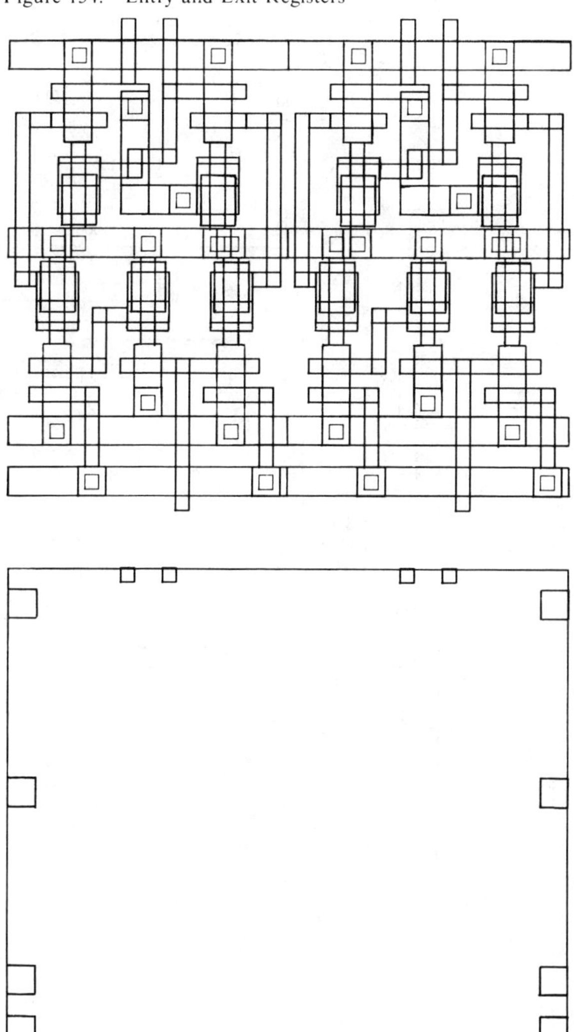

Figure 155. Logical Not

Figure 156. Selector

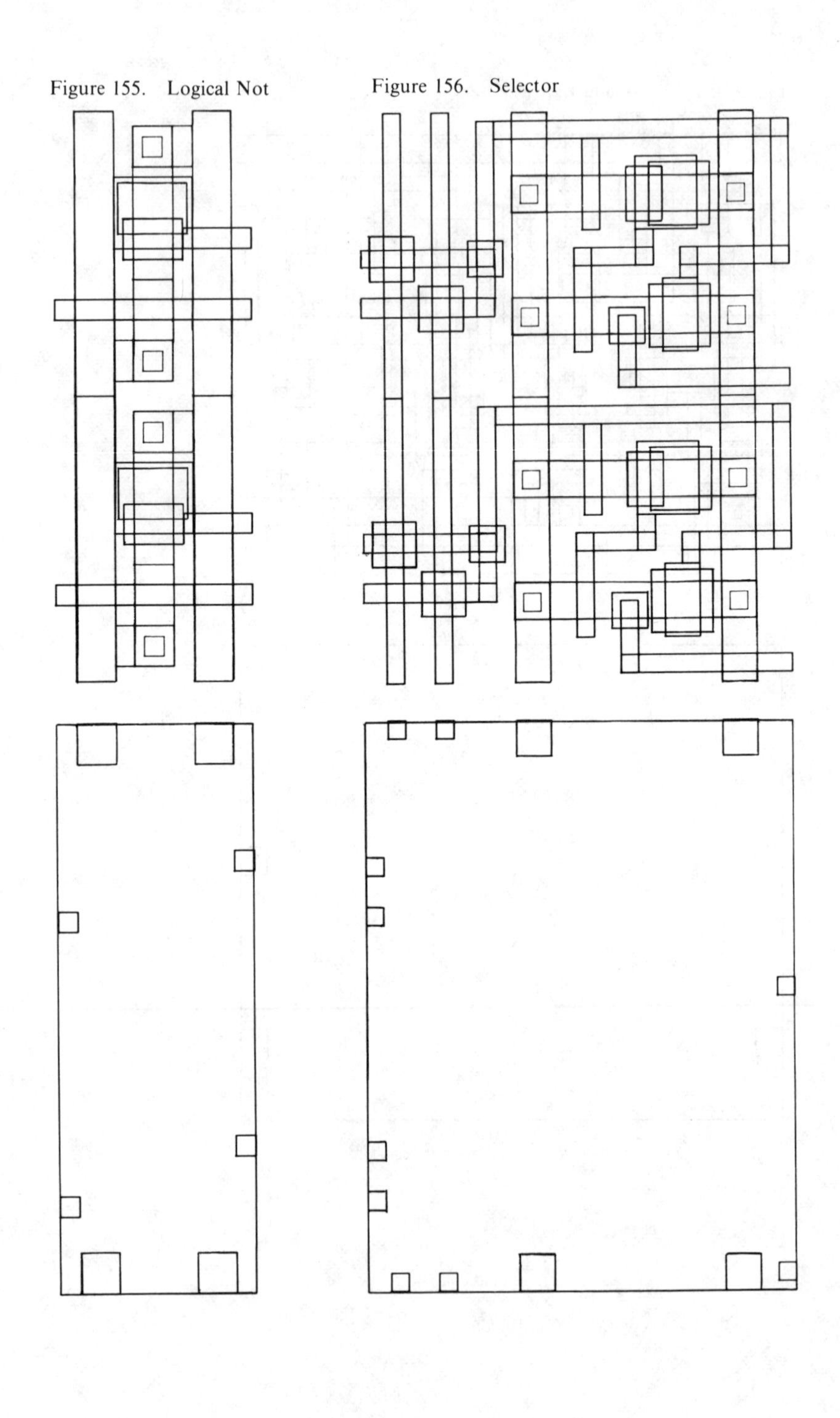

Notes

Chapter 2

1. Wires can be created on any of the three layers: poly, n-channel diffusion, and metal. Since the intersection of poly and diffusion forms a transistor, the use of these layers for interconnect is somewhat constrained. Hence, one-half layer is "missing" with respect to wiring.

Chapter 3

1. Although the term "specification" is used throughout this work, it is used colloquially rather than formally, since formal semantics have not been defined for d-n.

Chapter 4

1. d-n-defined functions are *not* functions in the mathematical sense. They are really hardware subroutines.
2. The physical description of an encapsulation is captured in its floor plan or physical diagram. Floor plans will be discussed in the section on translation.
3. The collective behavior of the receptor interfaces constitutes the visible or observable behavior of the subsystem.
4. The description of a defined datatype has a physical diagram as well. The physical diagram will be discussed in the translation section of this chapter.
5. A value is any symbol (e.g., a character, integer) which can be suitably encoded for processing by the d-tech-db primitives. No other assumptions are made.
6. An operator just calls another defined function and a primitive datatype function invokes some operation within a storage instance which does all the computational work.
7. d does not propose a solution for the placement and routing problem. Since a real CAD system based on d must address this problem, provisions are made for the production of a point to point wire list.

Chapter 7

1. In an earlier, related effort, an editor for a graphical programming language was constructed which used both a keyboard and tablet for user interaction [Dav80]. Users indicated that too much time was spent moving their hands between the keyboard and the tablet.

2. Unix is a trademark of Western Electric.

Chapter 8

1. The representational datatype model strongly resembles the cognitive model for design and functional reasoning as developed by Freeman and Newell [Fre71].

Bibliography

[Ayr83] Ronald F. Ayres. *VLSI silicon compilation and the art of automatic microchip design.* Prentice-Hall Inc., Englewood Cliffs, NJ, 1983.

[Bak80a] Clark M. Baker and Chris Terman. Tools for verifying integrated circuit designs. *Lambda I,* 3 (Fourth Quarter 1980).

[Bak80b] Clark M. Baker. Artwork analysis tools for VLSI circuits. Master's thesis, MIT, May 1980.

[Bar73] Mario R. Barbacci. Automated exploration of the design space for register transfer (RT) systems. Ph.D. thesis, Carnegie-Mellon University, November 1973.

[Bar74] Mario R. Barbacci and Daniel P. Siewiorek. Some aspects of the symbolic manipulation of computer descriptions. Tech. Rept. CMU-CS-74, Carnegie-Mellon University, July 1974.

[Bar81] Mario R. Barbacci. Instruction Set Processor Specifications (ISPS): The notation and its application. IEEE Trans. on Computers, Vol. C-30, No. 1, January 1981.

[Bat81] John Batali, Neil Mayle, Howard Shrobe, Gerald Sussman, and Daniel Weise. The DPL/Daedalus design environment. In *VLSI 81,* Academic Press, New York, 1981.

[Bel72] C. Gordon Bell, John Grason, and Allen Newell. *Designing computers and digital systems.* Digital Press, Burlington, MA, 1972.

[Bla81] Kyle M. Black and P. Kent Hardage. Advanced Symbolic Artwork Preparation (ASAP). *Hewlett Packard Journal 32,* 6 (June 1981).

[Bro83] Harold Brown, Christopher Tong, and Gordon Foyster. Palladio: An exploratory environment for circuit design. Computer, IEEE Computer Society, Vol. 16, No. 12, pp. 41-56.

[Bry80a] Randal E. Bryant. An algorithm for MOS logic simulation. *Lambda I,* 3 (Fourth Quarter 1980).

[Bry80b] Randal E. Bryant. MOSSIM: A logic-level simulator for MOS LSI user's manual. Tech. Rept. VLSI Memo No. 80-21, MIT, July 1980.

[Cap81] Peter R. Cappello and Kenneth Steiglitz. Digital signal processing applications of systolic algorithms. VLSI systems and computations, Rockville, MD, 1981.

[Car81] Tony M. Carter and Lee A. Hollaar. Implementation of asynchronous control unit state machines in integrated circuits using the Stored Logic Array (SLA). Tech. Rept. VLSI group technical memo, Computer Science Department, University of Utah, July 1981.

[Coo84] William A. Cook. TN: A structural description facility for hardware. Master's thesis, Case Western Reserve University, Cleveland, OH, 1984.

[Cla80] James H. Clark. Structuring a VLSI system architecture. *Lambda I,* 2 (Second quarter 1980).

[Dar80] John A. Darringer and William H. Joyner, Jr. A new look at logic synthesis. Tech. Rept. RC 8268, IBM Thomas J. Watson Research Center, May 1980.

[Dav79] Alan L. Davis. A data driven architecture suitable for VLSI implementation. Proceedings of the 1st Caltech Conference on VLSI, Caltech, January 1979.

[DavPC] Alan L. Davis. Personal communications. EARL is a manual wire editor under development at Caltech.

[Dav80] Alan L. Davis and Paul J. Drongowski. Dataflow computers: A tutorial and survey. Tech. Rept. UUCS-80-109, Computer Science Department, University of Utah, July 1980.

[Dav83] Randall Davis and Howard Shrobe. Representing structure and behavior of digital hardware. Computer, IEEE Computer Society, Vol. 16, No. 10, October 1983, pp. 75-82.

[Dir82] Steve W. Director, et al. A design methodology and computer aids for digital VLSI systems. Computer Science Research Review, Carnegie-Mellon University, 1982.

[Dol78] T.A. Dolotta, R.C. Haight, and J.R. Mashey. The programmer's workbench. *The Bell System Technical Journal 57,* 6 (July-August 1978), 2177-2200.

[Dro83] Paul J. Drongowski. System speed, space and power estimation using a higher level design notation. International Conference on Computer Design, IEEE Computer Society, October 1983, pp. 468-71.

[Dro84] Paul J. Drongowski. Functional simulation with the N.mPc system. VLSI Design, Vol. 5, No. 1, January 1984, pp. 76-77.

[Dro85] Paul J. Drongowski. Representation in CAD: Model and semantics. Submitted to the 1985 ACM Computer Science Conference, March 1985.

[Est77] Gerald Estrin. Modeling for synthesis—the gap between intent and behavior. Proceedings of the Symposium on Design Automation and Microprocessors, ACM SIGDA and IEEE Computer Society, Palo Alto, CA, February 1977, pp. 54-59.

[Fai80] Douglas Fairbairn and James Rowson. An interactive layout system. In *An introduction to VLSI systems,* Addison Wesley, Reading, MA, 1980.

[Fen81] Steven J. Fenves and Mary Lou Maher. The use of artificial intelligence techniques in preliminary structural design. Carnegie-Mellon University Design Research Center, Report DRC-12-04-81, August 1981.

[FraFE] Edward H. Frank. FETS: Fast Eddie's Timing Simulator. VLSI document, Carnegie-Mellon University.

[Fra81] Edward H. Frank and Robert F. Sproull. Testing and debugging custom integrated circuits. Tech. Rept. CMU-CS-81-105, Department of Computer Science, Carnegie-Mellon University, February 1981.

[Fre71] P. Freeman and A. Newell. A model for functional reasoning in design. Tech. Rept. CMU-CS-71-107, Carnegie-Mellon University, May 1971.

[Gra83] John J. Granacki and Alice C. Parker. The effect of register-transfer design tradeoffs on chip area and performance. Proceedings of the 20th Design Automation Conference, IEEE Computer Society and ACM SIGDA, June 1983, pp. 419-24.

[GraCTa] J.P. Gray. A VLSI design philosophy and support software. Caltech SSP File #3240.

[GraCTb] J.P. Gray and I. Buchanan. Models for structured IC design. Caltech SSP File #3230.

[Gut77] John Guttag. Abstract data types and the development of data structures. *CACM 20,* 6 (June 1977).

[Haf83] Louis J. Hafer and Alice C. Parker. A formal method for the specification, analysis and design of register-transfer level digital logic. IEEE Transactions on Computer-Aided Design, Vol. CAD-2, No. 1, January 1983, pp. 4-18.

[Har80] Kent Hardage. ASAP: Advanced Symbolic Artwork Preparation. *Lambda I,* 3 (Fourth Quarter 1980).

[Hen81] John Hennessy. SLIM: A simulation and implementation language. *Lambda II,* 2 (Second quarter 1981).

[Hil73] Frederick J. Hill and Gerald R. Peterson. *Digital systems: Hardware organization and design.* John Wiley & Sons, New York, NY, 1973.

[HollE] Lee A. Hollaar. Direct implementation of asynchronous control units. Submitted to IEEE Transactions on Computers.

[Hor83] Paul W. Horstmann. Expert systems and logic programming for CAD. VLSI Design, Vol. IV, No. 7, November 1983, pp. 37-46.

[JohCT] David Johannsen. Bristle blocks: A silicon compiler. Caltech Display File #2587.

[Joh79] David Johannsen. Bristle blocks: A silicon compiler. Proceedings of the 16th Design Automation Conference, June 1979.

[Joh82] Stephen C. Johnson. Hierarchical design validation based on rectangles. MIT Conference on Advance Research in VLSI, MIT, January 1982.

[Kat83] Randy H. Katz. Managing the chip design database. Computer, IEEE Computer Society, Vol. 16, No. 12, December 1983, pp. 26-36.

[Kel82] Kenneth H. Keller and A. Richard Newton. KIC2: A low-cost, interactive editor for integrated circuit design. Digest of Papers, Spring CompCon '82, IEEE Computer Society, February 1982.

[Knu69] Donald E. Knuth. *The art of computer programing.* Volume 1: *Fundamental algorithms.* Addison-Wesley, Reading, MA, 1969.

[Kow83] T.J. Kowalski and D.E. Thomas. The VLSI design automation assistant: Prototype system. Proceedings of the 20th Design Automation Conference, IEEE Computer Society and ACM SIGDA, June 1983, pp. 479-83.

[Kra83] Joseph C. Krauskopf. Control modules for asynchronous VLSI designs. Master's thesis, Case Western Reserve University, Cleveland, OH, 1983.

[Kun80] H.T. Kung. Special purpose devices for signal and image processing: An opportunity in VLSI. Tech. Rept. CMU-CS-80-132, Carnegie-Mellon University, July 1980.

[Kun81] H.T. Kung and S.W. Song. A systolic 2-D convolution chip. Tech. Rept. CMU-CS-81-110, Carnegie-Mellon University, March 1981.

[LaP80] Andrea S. LaPaugh. Algorithms for integrated circuit layout: An analytical approach. Ph.D. thesis, MIT, November 1980.

[Lat79] William Lattin. VLSI design methodology: the problem of the 80's for microprocessor design. Proceedings of the 1st Caltech Conference on VLSI, Caltech, January 1979.

[Lat81] William W. Lattin, John A. Bayliss, David L. Budde, Justin R. Rattner, and William S. Richardson. A methodology for VLSI chip design. *Lambda II,* 2 (Second quarter 1981).

[Lei81] Charles E. Leiserson and Ron Y. Pinter. Optimal placement for river routing. In *VLSI systems and computations,* Computer Science Press, Rockville, MD, 1981.

[Lip82] Richard J. Lipton, Robert Sedgewick, and Jacobo Valdes. Programming aspects of VLSI. Principles of Programming Languages, SIGPLAN, February 1982.

[LocCT] Bart Locanthi. LAP: A Simula package for IC layout. Caltech Display File #1862.

[Lyo81] Richard F. Lyon. Simplified design rules for VLSI layout. *Lambda II,* 1 (First Quarter 1981).

[Mat82] R. Mathews and J. Newkirk. A target language for silicon compilers. Digest of Papers, Spring CompCon '82, IEEE Computer Society, February 1982.

[MaC82] David W. McSweeney. Timing verification of VLSI logic circuits. MIT Conference on Advance Research in VLSI, MIT, January 1982.

[MaW78] T.M. McWilliams and L.C. Widdoes, Jr. SCALD: Structured Computer-Aided Logic Design. Proceedings of the 15th Design Automation Conference, June 1978.

[Mea80] Carver Mead and Lynn Conway. *An introduction to VLSI systems.* Addison Wesley, Reading, MA, 1980.

[Mos81] R.C. Mosteller. REST: A leaf cell design system. In *VLSI 81,* Academic Press, New York, 1981.

[Mud81] Craig Mudge. VLSI chip design at the crossroads. In *VLSI 81,* Academic Press, New York, 1981.

[New82] Martin E. Newell and Daniel T. Fitzpatrick. Exploiting structure in integrated circuit design analysis. MIT Conference on Advance Research in VLSI, MIT, January 1982.

[Not80] R. Noto, et al. Automated design procedures for VLSI. Research and Development Technical Report to the U.S. Army Electronics Research and Development Command, Fort Monmouth, NJ, February 1980.

[Ous81] John K. Ousterhout. Caesar: An interactive editor for VLSI layouts. *Lambda II*, 4 (Fourth quarter 1981).

[Ous82] John K. Ousterhout. Caesar: An interactive editor for VLSI layout. Digest of Papers, Spring CompCon '82, IEEE Computer Society, February 1982.

[Pak72] Sandra Pakin. *APL\360 Reference Manual, 2nd Edition.* Science Research Associates, Chicago, IL, 1972.

[Par79] Frederic I. Parke. An introduction to the N.mPc design environment. Proceedings of the 16th Design Automation Conference, ACM SIGDA, San Diego, CA, June 1979, pp. 513-19.

[Par79a] Alice C. Parker and Louis Hafer. Automated synthesis of digital hardware. Tech. Rept. DRC-18-12-79, Carnegie-Mellon University, May 1979.

[Par79b] Alice C. Parker, Donald E. Thomas, and Steve Crocker. ISPS: A retrospective view. Tech. Rept. DRC-18-14-79, Carnegie-Mellon University, May 1979.

[PetCP] John Peterson. Size reduction of SLA circuits. Final class project in artificial intelligence.

[Pet82] Phil Petit. Chipmonk: An interactive VLSI layout tool. Digest of Papers, Spring CompCon '82, IEEE Computer Society, February 1981.

[Pin81] Ron Y. Pinter. Optimal routing in rectilinear channels. In *VLSI systems and computations,* Computer Science Press, Rockville, MD, 1981.

[Rem79] Martin Rem. Mathematical aspects of VLSI design. Proceedings of the 1st Caltech Conference on VLSI, Caltech, January 1979.

[Rem81] Martin Rem. The VLSI challenge: Complexity bridling. In *VLSI 81,* Academic Press, New York, 1981.

[Ros84] Charles W. Rose, Greg Ordy, and Paul J. Drongowski. N.mPc: A study in university-industry technology transfer. Design and Test of Computers, IEEE Computer Society, February 1984.

[Rot81] J. Paul Roth and A.L. Frisiani. Computer timing verification. Tech. Rept. RC 8335, IBM Thomas J. Watson Research Center, July 1981.

[Row80] James A Rowson. *Understanding hierarchical design.* Ph.D. thesis, Caltech, April 1980.

[San78] E. Sandewall. Programming in the interactive environment: The LISP experience. *ACM Computing Surveys 10,* 1 (March 1978).

[San80] Alberto L. Sangiovanni-Vincentelli and N.B. Guy Rabbat. Techniques for the time-domain analysis of LSI circuits. Tech. Rept. RC 8351, IBM Thomas J. Watson Research Center, July 1980.

[Sas82] Sarma Sastry and Steve Klein. Plates: A metric-free VLSI layout language. MIT Conference on Advance Research in VLSI, MIT, January 1982.

[Sei79] Charles L. Seitz. Self-timed VLSI systems. Proceedings of the 1st Caltech Conference on VLSI, Caltech, January 1979.

[Sei80] Charles L. Seitz. System timing. In *Introduction to VLSI systems,* Addison Wesley, Reading, MA, 1980.

[ShiIP] Jerry Shields. Present and future CAD tools for VLSI from Computervision. Invited presentation. Computer Science Department, University of Utah.

[Shr82a] Howard E. Shrobe. The data path generator. Digest of Papers, Spring CompCon '82, IEEE Computer Society, February 1982.

[Shr82b] Howard E. Shrobe. The data path generator. MIT Conference on Advance Research in VLSI, MIT, January 1982.

[Sis82] Jeffrey M. Siskind, Jay R. Southard, and Kenneth W. Crouch. Generating custom high performance VLSI designs from succinct algorithmic descriptions. MIT Conference on Advance Research in VLSI, MIT, January 1982.

[Smi83] David C. Smith, et al. VAGUA: The variable geometry automated universal array layout system. Proceedings of the 20th Design Automation Conference, IEEE Computer Society and ACM SIGDA, June 1983, pp. 425-29.

[Smi81a] Kent F. Smith. Implementation of SLA's in nMOS technology. In *VLSI 81,* Academic Press, New York, 1981.

[Smi81b] Kent F. Smith, Tony M. Carter, and Charles E. Hunt. The CMOS SLA implementation and SLA program structures. In *VLSI systems and computations,* Computer Science Press, Rockville, MD, 1981.

[SmiPC] Kent Smith. Personal communications.

[Sno78] Edward A. Snow III. *Automation of module set independent register transfer level design.* Ph.D. thesis, Carnegie-Mellon University, April 1978.

[Sor83] Michael J. Sorens. An adaptable graph editor for a computer aided design system. Master's thesis, Case Western Reserve University, Cleveland, OH, 1983.

[Spi81] *SPICE User's Manual.* University of Utah VLSI Group, 1981.

[Spr80] Robert F. Sproull and Richard F. Lyon. Introduction to VLSI systems (The Caltech Intermediate Form for LSI Layout Description). In *Introduction to VLSI systems,* Addison Wesley, Reading, MA, 1980.

[Sta81] Editorial staff. VLSI design and tool development at DEC. *Lambda II,* 1 (First Quarter 1981).

[Sta81] Richard M. Stallman. EMACS: The extensible, customizable self-documenting display editor. Proceedings of the ACM SIGPLAN SIGOA Symposium on Text Manipulation, ACM SIGPLAN and SIGOA, June 1981.

[Ste82] Mark Stefik, Daniel G. Bobrow, Alan Bell, Harold Brown, Lynn Conway, and Christopher Tong. The partitioning of concerns in digital systems design. Proceedings of the Conference on Advance Research in VLSI, MIT, January 1982, pp. 43-52.

[Sub83a] P.A. Subrahmanyam. Overview of a conceptual and formal basis for an automatable high level design paradigm for integrated systems. Proc. ICCD'83, IEEE Computer Society, Port Chester, NY, 1983, pp. 647-51.

[Sub83b] P.A. Subrahmanyam. Synthesizing VLSI circuits from behavioral specifications: A very high level silicon compiler and its theoretical basis. *VLSI 83,* F. Anceau (editor), North Holland, Amsterdam, 1983.

[Sut77] Ivan E. Sutherland and Carver A. Mead. Microelectronics and computer science. *Scientific American 237,* 3 (September 1977).

[Suz82] Norihisa Suzuki and Rod Burstall. Sakura: A VLSI modelling language. MIT Conference on Advance Research in VLSI, MIT, January 1982.

[Tho83a] Donald E. Thomas and John A. Nestor. Designing and implementing a multilevel design representation with simulation applications. IEEE on Computer-Aided Design, Vol. CAD-2, No. 3, July 1983, pp. 135-45.

[Tho83b] Donald E. Thomas, et al. Automatic data path synthesis. Computer, IEEE Computer Society, Vol. 16, No. 12, December 1983, pp. 59-70.

[Tri80] S. Trimberger. The proposed Sticks standard. Tech. Rept. SSP Memo #3487, Caltech, 1980.

[Tur79] David A. Turner. SASL language manual. Tech. Rept., Computer Laboratory, University of Kent, July 1979.

[Van79] William M. vanCleemput. Hierarchical design for VLSI: Problems and advantages. Proceedings of the 1st Caltech Conference on VLSI, Caltech, January 1979.

[Van82] Willaim M. vanCleemput. CAD tools for custom integrated circuit design. Digest of Papers, Spring CompCon '82, IEEE Computer Society, February 1982.

[Wei82] Daniel L. Weinreb. High performance personal computation for VLSI CAD. Digest of Papers, Spring CompCon '82, IEEE Computer Society, February 1982.

[Wei81] Uri Weiser and Alan L. Davis. A wavefront notation for VLSI array design. In *VLSI systems and computations,* Computer Science Press, Rockville, MD, 1981.

[Whi81a] Telle Whitney. A hierarchical design-rule checking algorithm. *Lambda II,* 1 (First Quarter 1981).

[Whi81b] Telle Whitney. A hierarchical design analysis front end. In *VLSI 81,* Academic Press, New York, 1981.

[Wil77] J. Williams. Sticks—a new approach to LSI design. Master's thesis, MIT, 1977.

Index